Practical Industrial Safety, Risk Asses and Shutdown Systems for Industry

Titles in the series

Practical Cleanrooms: Technologies and Facilities (David Conway)

Practical Data Acquisition for Instrumentation and Control Systems (John Park, Steve Mackay)

Practical Data Communications for Instrumentation and Control (Steve Mackay, Edwin Wright, John Park)

Practical Digital Signal Processing for Engineers and Technicians (Edmund Lai)

Practical Electrical Network Automation and Communication Systems (Cobus Strauss)

Practical Embedded Controllers (John Park)

Practical Fiber Optics (David Bailey, Edwin Wright)

Practical Industrial Data Networks: Design, Installation and Troubleshooting (Steve Mackay, Edwin Wright, John Park, Deon Reynders)

Practical Industrial Safety, Risk Assessment and Shutdown Systems for Instrumentation and Control (Dave Macdonald)

Practical Modern SCADA Protocols: DNP3, 60870.5 and Related Systems (Gordon Clarke, Deon Reynders)

Practical Radio Engineering and Telemetry for Industry (David Bailey)

Practical SCADA for Industry (David Bailey, Edwin Wright)

Practical TCP/IP and Ethernet Networking (Deon Reynders, Edwin Wright)

Practical Variable Speed Drives and Power Electronics (Malcolm Barnes)

Practical Industrial Safety, Risk Assessment and Shutdown Systems for Industry

Dave Macdonald BSc(Eng)

AMSTERDAM • BOSTON • HEIDELBERG • LONDON • NEW YORK • OXFORD
PARIS • SAN DIEGO • SAN FRANCISCO • SINGAPORE • SYDNEY • TOKYO

Newnes is an imprint of Elsevier

Newnes

Newnes
An imprint of Elsevier
Linacre House, Jordan Hill, Oxford OX2 8DP
200 Wheeler Road, Burlington, MA 01803

First published 2004

British Library Cataloguing in Publication Data
A catalogue record for this book is available from the British Library

ISBN   07506 58045

For information on all Newnes publications,
visit our website at www.newnespress.com

Typeset and Edited by Vivek Mehra, Mumbai, India
(vivekmehra@tatanova.com)

Transferred to digital printing in 2010

Contents

4   Safety requirements specifications   108

Preface

Most of today's computer controlled industrial processes involve large amounts of energy and have the potential for devastating accidents. Reliable, well-engineered safety systems are essential for protection against destruction and loss of life.

This book is an intensive practical and valuable exposure to the most vital, up-to-date information and practical know-how to enable you to participate in hazard studies and specify, design, install and operate the safety and emergency shutdown systems in your plant, using international safety practices.

This book will provide you with a broad understanding of the latest safety instrumentation practices and their applications to functional safety in manufacturing and process industries. This book could save your business a fortune in possible downtime and financial loss.

The objectives of the book are to:

- Expand your practical knowledge in the application of safety instrumented systems (SIS) as applied to industrial processes
- Provide you with the knowledge of the latest standards dealing with each stage of the safety life cycle from the initial evaluation of hazards to the detailed engineering and maintenance of safety instrumented systems
- Give you the ability to plan hazard and risk assessment studies, then design, implement and maintain the safety systems to ensure high reliability
- Assist your company to implement functional safety measures to international standards

There are least six practical exercises to give you the hands-on experience you will need to implement and support hazard studies; perform reliability evaluations; specify requirements; design, plan and install reliable safety and emergency shutdown systems in your business.

Although a basic understanding of electrical engineering principles is essential, even those with a superficial knowledge will substantially benefit by reading this book.

In particular, if you work in any of the following areas, you will benefit from reading this book:

- Instrumentation and control engineers and technicians
- Design, installation and maintenance engineers and technicians in the process industries
- Managers and sales professionals employed by end users
- Systems integrators
- Systems consultants
- Consulting electrical engineers
- Plant engineers and instrument technicians
- Operations technicians
- Electrical maintenance technicians and supervisors
- Instrumentation and control system engineers
- Process control engineers
- Mechanical engineers

The structure of the book is as follows.

Chapter 1: Introduction. A review of the fundamentals in safety instrumentation focussing on a discussion on hazards and risks, safety systems engineering, and introduction to the IEC 61508 and ISA S84 standards. A concluding review of the safety life cycle model and its phases.

Chapter 2: Hazards and risk reduction. An examination of basic hazards, the chemical process, hazards studies, the IEC model, protection layers, risk reduction and classification and the important concept of the safety integrity level (SIL).

Chapter 3: Hazard studies. A review of the outline of methodologies for hazard studies 1, 2 and 3.

Chapter 4: Safety requirements specifications. A discussion and guide to preparing a safety requirements specification (SRS).

Chapter 5: Technology choices and the conceptual design stage. An examination of how to get the concepts right for the specific application and choosing the right type of equipment for the job, not the particular vendor but at least the right architecture for the logic solver system and the right arrangement of sensors and actuators to give the quality of system required by the SRS.

Chapter 6: Basic reliability analysis applied to safety systems. This discusses the task of measuring or evaluating the SIS design for its overall safety integrity.

Chapter 7: Safety in field instruments and devices. This chapter examines the range of instrumentation design techniques that have accumulated in the industry through experience that began a long time before the days of PES and the high performance logic solvers.

Chapter 8: Engineering the safety system: hardware. An examination of two aspects of engineering work for building an SIS. Firstly there is a look at some aspects of project engineering management and secondly some basic engineering practices.

Chapter 9: Engineering the application software. Guidance is provided here on how to deal with the application software stages of an SIS project with an examination of some of the basic concepts and requirements that have been introduced in recent years to try to overcome the major concerns that have arisen over the use of software in safety applications.

Chapter 10: Overall planning: IEC Phases 6, 7 and 8. A brief look at the planning boxes marked in on the IEC safety life cycle.

Chapter 11: Installation and commissioning (IEC phase 12). This chapter tracks the safety system from its building stage through factory acceptance testing, delivery and installation and into final testing for handover to the operating team.

Chapter 12: Validation, operations and management of change (IEC phases 13, 14 and 15). A discussion on validation, operations and maintenance.

Chapter 13: Justification for a safety instrumented system. In practice engineers and managers have to make choices on the type, quality, and costs of the safety solutions available within the constraints imposed by the essential safety requirements. This is discussed in detail in this chapter.

1

Introduction

1.1   Definition of safety instrumentation

What is safety instrumentation?

Here is a typical definition.

(Origin: UK Health and Safety Executive: 'Out of Control')

'Safety instrumented systems are designed to respond to conditions of a plant that may be hazardous in themselves or if no action were taken could eventually give rise to a hazard. They must generate the correct outputs to prevent the hazard or mitigate the consequences'

Abbreviation: The acronym SIS means 'safety instrumented system'. We probably all know the subject by other names because of the different ways in which these systems have been applied. Here are some of the other names in use:

- Trip and alarm system
- Emergency shutdown system
- Safety shutdown system
- Safety interlock system
- Safety related system (more general term for any system that maintains a safe state for EUC)

Fig1.1 defines the SIS as bounded by sensors, logic solver and actuators with associated interfaces to users and the basic process control system. We are talking about automatic control systems or devices that will protect persons, plant equipment or the environment against harm that may arise from specified hazardous conditions.

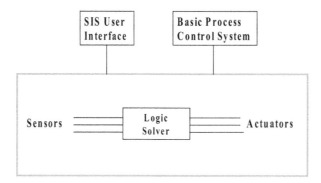

Figure 1.1
Definition of a safety instrumented system

We are talking about automatic control systems or devices that will protect persons, plant equipment or the environment against harm that may arise from specified hazardous conditions.

1.2   What is this book about?

This book is about instrumentation and control systems to support:

- The safety of people in their workplaces
- Protecting the environment against damage from industrial accidents
- Protecting businesses against serious losses from damage to plant and machinery
- Creating awareness of the good practices available for the delivery of effective safety instrumented systems
- Providing basic training in well established techniques for engineering of safety systems
- Assisting engineers and technicians to support and participate in the safety systems activities at their work with a good background knowledge of the subject
- Being aware of what can go wrong and how to avoid it

1.3   Why is this book necessary?

- Safety systems are reaching wider fields of application
- Safety requires a multidiscipline approach
- New standards and new practices have emerged

There have been some steadily developing trends in the last 10 years which have moved the subject of so-called functional safety from a specialized domain of a few engineers into the broader engineering and manufacturing fields.

Basically, there is a need for a book to allow engineers and technicians to be aware of what is established practice in the safety instrumentation field without having to become specialists. After all it is the technicians who have to service and maintain the safety systems and they are entitled to know about the best available practices.

This book is also intended to be useful for:

- Project engineers and designers who may be involved in completely new projects or in the modification/upgrading of existing plants
- Engineers involved in the development of packaged processing plants or major equipment items where automatic protection systems may be needed
- Engineers and technicians working for instrumentation and control system suppliers

1.4 Contents of the book

The subjects in this book cover the '*life cycle*' of safety protection from the initial studies and requirements stages through to the operation and support of the finished systems, i.e.

- Identification of hazards and specification of the protection requirements
- Technology choices
- Engineering of the protection systems
- Operations and maintenance including control of changes

This subject is well supplied with specialized terms and abbreviations, which can be daunting and confusing. We have attempted to capture as many as possible in a glossary. This is located at the back of the book.

Reference book: Acknowledgments are given to the authors of the following book for many helpful features in their book that have been of assistance in the preparation of this particular book. Details of this book are as follows:

Title: *Safety Shutdown Systems: Design, Analysis and Justification*
By: Paul Gruhn and Harry Cheddie
Published by: Instrument Society of America, 1998. ISBN 1-5517-665-1
Available from ISA Bookstore website: www.isa.org

1.5 Introduction to hazards and risks

The first part of the book is all about the identification of hazards and the reduction of the risks they present.

What is a hazard and what is a risk?

A hazard is '*an inherent physical or chemical characteristic that has the potential for causing harm to people, property, or the environment*'

In chemical processes: '*It is the combination of a hazardous material, an operating environment, and certain unplanned events that could result in an accident.*'

Risk: '*Risk is usually defined as the combination of the severity and probability of an event. In other words, how often can it happen and how bad is it when it does? Risk can be evaluated qualitatively or quantitatively*'

Roughly: RISK = FREQUENCY × CONSEQUENCE OF HAZARD

Consider the risk on a cricket field.

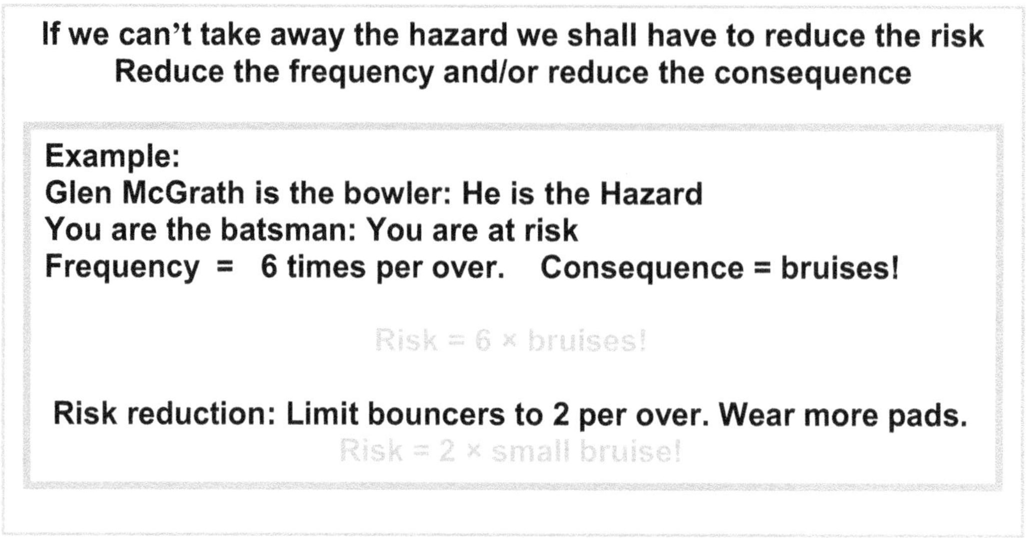

If we can't take away the hazard we shall have to reduce the risk
Reduce the frequency and/or reduce the consequence
Example:
Glen McGrath is the bowler: He is the Hazard
You are the batsman: You are at risk
Frequency = 6 times per over. Consequence = bruises!
Risk = 6 × bruises!
Risk reduction: Limit bouncers to 2 per over. Wear more pads.
Risk = 2 × small bruise!

Figure 1.2
Risk reduction: the fast bowler

1.5.1 Risk reduction

The reduction of risk is the job of protection measures. In some cases this will be an alternative way of doing things or it can be a protection system such as a safety instrumented system. When we set out designing a protection system we have to decide how good it must be. We need to decide how much risk reduction we need (and this can be one of the hardest things to agree on). The target is to reduce the risk from the unacceptable to at least the tolerable. This principle has a fundamental impact on the way we have to design a safety system as shown in the following diagram.

Hazard Identified
Risk
Estimated/Calculated
Tolerable Risk
Established
Risk Reduction
Requirement
Safety Function
Defined

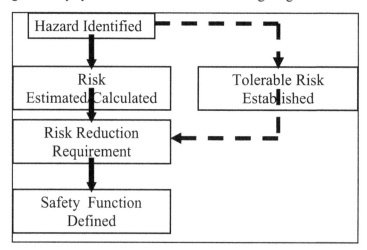

Figure 1.3
Risk reduction: design principles

The concept of tolerable risk is illustrated by the following diagram showing what is known as the principle of ALARP.

ALARP boundaries for individual risks: Typical values.

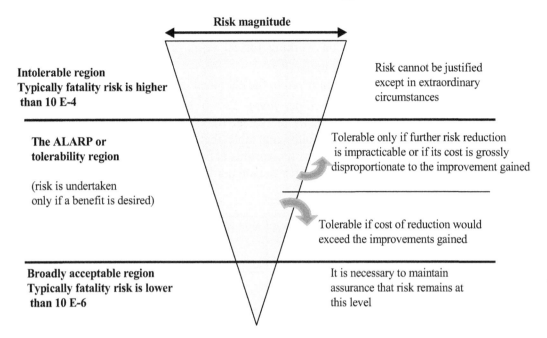

Figure 1.4
Principle of ALARP

The ALARP (as low as reasonably practicable) principle recognizes that there are three broad categories of risks:

- Negligible Risk: broadly accepted by most people as they go about their everyday lives, these would include the risk of being struck by lightning or of having brake failure in a car.
- Tolerable risk: We would rather not have the risk but it is tolerable in view of the benefits obtained by accepting it. The cost in inconvenience or in money is balanced against the scale of risk and a compromise is accepted. This would apply to traveling in a car, we accept that accidents happen but we do our best to minimize our chances of disaster. Does it apply to Bungee jumping?
- Unacceptable risk: The risk level is so high that we are not prepared to tolerate it. The losses far outweigh any possible benefits in the situation.

Essentially this principle guides the hazard analysis participants into setting tolerable risk targets for a hazardous situation. This is the first step in setting up a standard of performance for any safety system.

1.6   Fatal accident rate (FAR)

This is one method of setting a tolerable risk level. If a design team is prepared to define what is considered to be a target fatal accident rate for a particular situation it becomes possible to define a numerical value for the tolerable risk. Whilst it seems a bit brutal to set such targets the reality is that certain industries have historical norms and also have targets for improving those statistical results.

The generally accepted basis for quoting FAR figures is the number of fatalities per one hundred million hours of exposure. This may be taken as the fatalities per 10^8 worked hours at a site or in an activity but if the exposure is limited to less than all the time at work this must be taken into account.

Very roughly 1 person working for 50 years or 50 people working for 1 year will accumulate 10^5 working hours

If 50 000 people are employed in the chemical industries there will be an average of:

50 000 × 2000 hrs worked per year = $1 × 10^8$ hrs worked per year. If the same industry recorded an FAR of 4 it means an average of 4 fatalities per year has occurred.

You can see from the following table that this scale of measurement allows some comparisons to be made between various activities. Another scale of measurement is the probability of a fatal accident per person per year for a particular activity.

Activity	FAR per 10^8	Individual risk of death per person per year × 10^{-4}
Travel		
Air		0.02
Train	3–5	0.03
Bus	4	2
Car	50–60	2
Occupation		
Chemical industry	4	0.5
Manufacturing	8	
Shipping	8	9
Coal mining	10	2
Agriculture	10	
Boxing	20 000	
Rock climbing	4 000	1.4
Staying at home	1–4	
Living at 75 (based on simple calculation of hr/lifetime)	152	133

Table 1.1
Individual risk and fatal accident rates based on UK data

FAR can be used as basis for setting the tolerable rate of occurrence for a hazardous event. For example:

Suppose a plant has an average of 5 persons on site at all times and suppose that 1 explosion event is likely to cause 1 person to be killed. The site FAR has been set at 2.0 × 10^{-8}/hr. We can calculate the minimum average period between explosions that could be regarded as tolerable, as follows:

Fatality rate per year = (FAR/hr) × (hours exposed/yr)

$$= (2 × 10^{-8}) × (5 × 8760)$$

$$= 8.76 × 10^{-4}$$

Avg. years per explosion = $1/8.76 × 10^{-4}$ = 1140 year

Note: If there are N separate sources of explosion of the same type the period for each source will be: $N \times 1140$ years. These figures will define the target risk frequencies for determining the scale of risk reduction needed from a safety system.

1.7 Overview of safety systems engineering (SSE)

The term safety systems engineering is used to describe the systematic approach to the design and management of safety instrumented systems.

1.7.1 Introduction

Safety systems engineering (SSE) comprises all the activities associated with the specification and design of systems to perform safety functions. SSE has become a discipline within the general field of engineering. Whenever there is a clear and obvious need for safety to be engineered into any activity it should be done properly and in a systematic manner.

1.7.2 What do we mean by safety functions?

We mean any function that specifically provides safety in any situation. E.g. a seat belt in a car, an air bag, a pressure relief valve on a boiler or an instrumented shutdown system. Thus an air bag has a safety function to prevent injury in the event of collision. The safety system of an air bag comprises the sensor, the release mechanism, the inflator and the bag itself.

1.7.3 Functional safety

The term 'functional safety' is a concept directed at the functioning of the safety device or safety system itself. It describes the aspect of safety that is associated with the functioning of any device or system that is intended to provide safety. The best description might be this one from the following journal article:

'Functional safety in the field of industrial automation' by Hartmut von Krosigk. *Computing and Control Engineering Journal* (UK IEE) Feb 2000.

'In order to achieve functional safety of a machine or a plant the safety related protective or control system must function correctly and, when a failure occurs, must behave in a defined manner so that the plant or machine remains in safe state or is brought into a safe state.'

Short form: *'Functional safety is that part of the overall safety of a plant that depends on the correct functioning of its safety related systems.'*

(Modified from IEC 61508 part 4.)

The next diagram shows how functional safety makes a contribution to overall safety.

Overall Safety is seen as part of overall safety

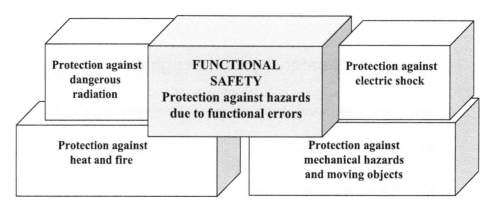

Figure 1.5
Overall safety

The well-known standards certification authority in Germany is TUV. Their website answers the question '*What is functional safety?*'

Random hardware faults or systematic design errors – e.g. in software – or human mistakes shall not result in a malfunction of a safety related unit/system with the potential consequence of:

- Injury or death of humans or
- Hazards to the environment or
- Loss of equipment or production

Then follows an explanation of the term '*unit/system*'; for example:

- A simple device as a gas burner control unit
- A large distributed computer system like emergency shutdown and fire & gas systems
- A field instrument
- The complete instrumented protective equipment of a plant

So we can conclude that *functional safety* is about the correct functioning of a unit or system designed to protect people and equipment from hazards.

1.8   Why be systematic?

Why be so formal? Why be systematic?
Critics might say...

- We don't need all these rules!
- Why not just use common sense?
- Whose job is it anyway?
- Make the contractor do it!

But now let's take a look at the problem.

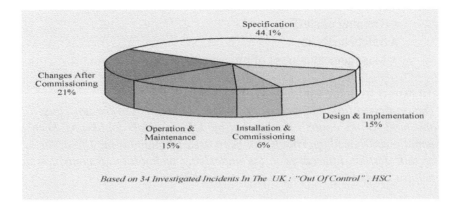

Figure 1.6
Causes of control system failures

Specification errors dominate the causes of accidents analyzed in the above survey.

1.8.1   UKHSE publication

One of the best advocates for a systematic approach to safety engineering is the UK Health and Safety Executive (HSE): Their publication: *'Out of Control'* is a very useful little book about *'Why control systems go wrong and how to prevent failure'* and it is the origin of the analysis we have just seen.

> Out of Control:
> Why control systems go wrong and how to prevent failure
> Published by UK Health and Safety Executive
> Contact: http://www.hse.gov.uk

This book not only provides extracts from the analyses of accidents but also explains with great clarity the need for a systematic approach to the engineering of functional safety. It also provides a valuable outline of the safety life cycle.

1.8.2   HSE summary

Some of the key points from the study are listed below:

Analysis of incidents

- Majority of incidents could have been anticipated if a systematic risk-based approach had been used throughout the life of the system
- Safety principles are independent of the technology
- Situations often missed through lack of systematic approach

Design problems

- Need to verify that the specification has been met
- Over dependence on single channel of safety
- Failure to verify software
- Poor consideration of human factors

Operational problems

- Training of staff
- Safety analysis
- Management control of procedures

(An extract from the summary is given below).

'*The analysis of the incidents shows that the majority were not caused by some subtle failure mode of the control system, but by defects which could have been anticipated if a systematic risk-based approach had been used throughout the life of the system. It is also clear that despite differences in the underlying technology of control systems, the safety principles needed to prevent failure remain the same.*'

Specification

'*The analysis shows that a significant percentage of the incidents can be attributed to inadequacies in the specification of the control system. This may have been due either to poor hazard analysis of the equipment under control, or to inadequate assessment of the impact of failure modes of the control system on the specification. Whatever the cause, situations which should have been identified are often missed because a systematic approach had not been used. It is difficult to incorporate the changes required to deal with the late identification of hazards after the design process has begun, and more difficult, (and expensive), to make such changes later in the life of the control system. It is preferable to expend resources eliminating a problem, than to expend resources in dealing with its effects.*'

Design

'*Close attention to detail is essential in the design of all safety-related control systems, whether they are simple hard-wired systems, or complex systems implemented by software. It is important that safety analysis techniques are used to ensure that the requirements in the specification are met, and that the foreseeable failure modes of the control system do not compromise that specification. Issues of concern, which have been identified, include an over-optimistic dependence on the safety integrity of single channel systems, failure to adequately verify software, and poor consideration of human factors. Good design can also eliminate, or at least reduce, the chance of error on the part of the operator or maintenance technician.*'

Maintenance and modification

'*The safety integrity of a well designed system can be severely impaired by inadequate operational procedures for carrying out the maintenance and modification of safety-related systems. Training of staff, inadequate safety analysis, inadequate testing, and inadequate management control of procedures were recurring themes of operational failures.*'

1.8.3 Conclusion: It pays to be systematic

Being systematic allows us to:

- Benefit from previously acquired knowledge and experience
- Minimize the chances of errors

- Demonstrates to others that we have done the job properly... they recognize our way of doing things as legitimate
- Makes it easier to compare one solution or problem with another and hence leads to generally accepted standards of protection
- Allows continuity between individuals and between different participants in any common venture – makes the safety system less dependent on any one individual
- Encourages the development of safety products that can be used by many
- Support regulatory supervision and compliance

1.8.4   Scope 1 of safety systems engineering

The next diagram shows how safety system engineering covers the whole life of an application. Quality assurance practices support the application at every stage.

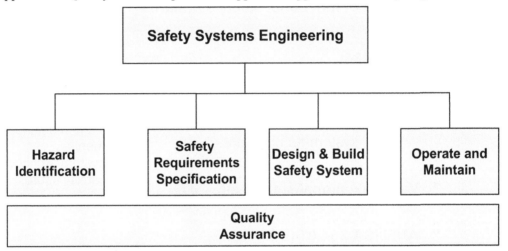

Figure 1.7
Scope of safety systems engineering

1.9   Introduction to standards: IEC 61508 and ISA S84

Up until the 1980s the management of safety in hazardous processes was left to the individual companies within the process industries. Responsible companies evolved sensible guidelines out of the knowledge that if they didn't take care of the problem they would be the nearest people to the explosion when it happened. The chemical industry for example was always aware that self-regulation would be better than rules imposed by a worried public through government action.

More recently, industry guidelines have matured into international standards and government regulators are seeing the potential benefits of asking companies and products to conform to what are becoming generally agreed standards. It's ironic that the better the standard the easier it becomes to enforce laws requiring conformance to that standard.

Here we take a look at how we have arrived at the point where new international standards are available. Then we look at the main standards to be used in this book.

1.9.1   Driving forces for management of safety

There are many reasons for wanting to improve the management of safety.

- We (the public) want to know that safety is properly organized
- Cost of accidents, catastrophes
- Rewards are high if the risk is low (Nuclear power)
- SHE Responsibilities of companies, designers and operators
- Legal requirements
- Complexities of processes and plants
- Hazards of multiple ownership
- Falling through the cracks. (Railways)
- Liabilities of owners, operators and designers
- Insurance risks and certification
- Programmable Electronic Systems (PES)

1.9.2 Evolution of functional safety standards

- **TUV (1984)**

- **DIN V 19250 / VDE V 0801 (Germany)**
  - Risk classification 1989
  - Safety system requirements

- **Various national standards**

- **ANSI/ISA S84.01 (USA) 1996**
  - Safety procedures
  - Safety life cycle

- **NFPA/UL1998**

- **OSHA (29 CFR 1910.119)**

- **UK HSE**

IEC 61508 98-2000
  - Overall safety life cycle
  - Safety plan/management
  - Safety integrity levels
  - Safety system diagnostic requirements
  - Safety system architectures and reliability figures

Courtesy: Honeywell SMS

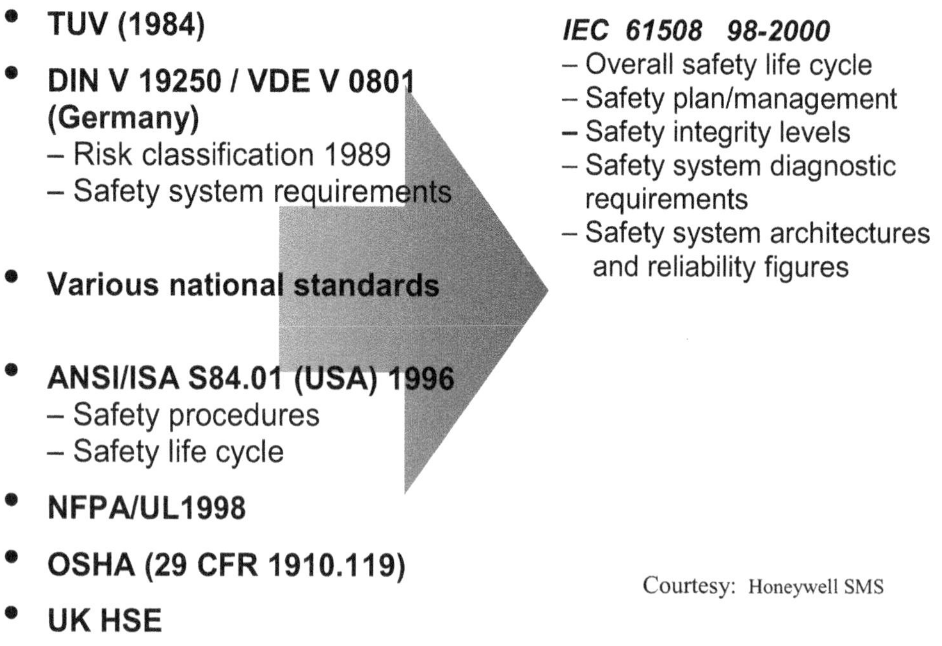

Figure 1.8
Evolution of functional safety standards

Programmable systems and network technologies have brought a new set of problems to functional safety systems. Software comes with new possibilities for performance failure due to program errors or untested combinations of coded instructions. Hence conventional precautions against defects in electrical hardware will not be sufficient to ensure reliability of a safety system.

Earlier design standards did not provide for such possibilities and hence they became obsolete.

Newer standards such as the German VDE 0801 and DIN 19250 emerged in the late 1980s to incorporate quality assurance grading for both hardware and software matched to the class of risk being handled. In the USA the ISA S84.01 standard was issued in 1995 for use in process industry applications including programmable systems. In the UK the

HSE promoted the drive for an international standard. These and many other factors have resulted in the issue of a new general standard for functional safety using electronic and programmable electronic equipment. The new standard issued by the IEC is IEC 61508 and it covers a wide range of activities and equipment associated with functional safety.

The newer standards bring a new approach to the management and design of functional safety systems. They try to avoid being prescriptive and specific because experience has shown that: '*A cookbook of preplanned solutions does not work.*'

The new approach is to set down a framework of good practices and limitations leaving the designers room to find appropriate solutions to individual applications.

1.9.3   Introducing standard IEC 61508

International Electrotechnical Commission

Title:

Functional safety of electrical/electronic/programmable electronic safety-related systems –

All Sections of IEC 61508 Now Published
Part 1:   General requirements
Part 2:   Requirements for electrical/electronic/programmable electronic systems
Part 3:   Software requirements
Part 4:   Definitions and abbreviations
Part 5:   Examples of methods for the determination of safety integrity levels
Part 6:   Guidelines on the application of parts 2, 3
Part 7:   Overview of techniques and measures
See Appendix 1 for Framework Diagram

Figure 1.9
Standard IEC 61508

This diagram shows the title of the standard and its 7 parts issued to date. An additional part 8 is in preparation, which will provide a further set of guidelines for the application of the standard.

1.9.4   Key elements of IEC 61508

- Management of functional safety
- Technical safety requirements
- Documentation
- Competence of persons

1.9.5   Features of IEC 61508

- Applies to safety systems using Electrical/Electronic/Programmable Electronic Systems (abbreviation: E/E/PES) e.g. Relays, PLCs, Instruments, Networks
- Considers all phases of the safety life cycle including software life cycle
- Designed to cater for rapidly developing technology
- Sets out a 'generic approach' for safety life cycle activities for E/E/PES

- Objective to 'facilitate the development of application sector standards'
- IEC 61511: process industry sector standard on the way

The standard is 'generic', i.e. it provides a generalized approach to the management and design of functional safety systems that can be applicable to any type of industry. It is intended for direct use in any project but it is also intended to be the basis for 'industry sector' standards. Hence, more specific industry sector standards will be expected to follow with alignment of their principles to the 'master standard'.

The IEC standard sets out procedures for managing and implementing a safety life cycle (abbr: SLC) of activities in support of a functional safety system. Hence, we can map the various parts of the standard on to our previous diagram of the safety life cycle as shown in the next diagram.

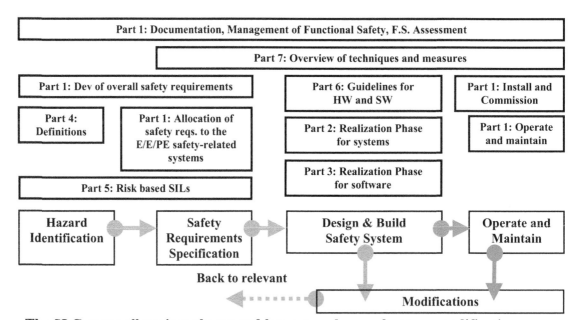

The SLC spans all project phases and has return loops whenever modifications

Figure 1.10
Framework of IEC 61508 relevant to SLC

1.9.6   Introducing Standard ANSI/S 84.01

Instrument Society of America

Title:
Application of Safety Instrumented Systems for the Process Industries

Sections of ISA S84.01
Clauses 1-11: Mandatory requirements
Clause 12:   Key differences from IEC 61508
Annexes A-E: Non mandatory (informative) technical information

Associated Document:
Draft Technical Report: 84.02 (ISA-dTR84.02)
Provides non mandatory technical guidance in Safety Integrity Levels

---

Figure 1.11
Standard ANSI/ISA S84.01 (USA) 1996

Features of ISA S84.01

- Applies to safety instrumented systems for the process industries
- Applies to safety systems using electrical/electronic/programmable electronic systems (abbr: E/E/PES)
- Defines safety life cycle activities for E/E/PES but excludes hazard definition steps associated with process engineering
- Objective: 'Intended for those who are involved with SIS in the areas of: design and manufacture of SIS products, selection and application installation, commissioning and pre-start-up acceptance test operation, maintenance, documentation and testing'

The ISA standard is a much less ambitious standard than IEC 61508 and it confines itself to the core instrument engineering activities relevant to process industries. It does not attempt to deal with the hazard study and risk definition phases of the safety life cycle.

1.9.7   Introducing Draft Standard IEC 61511

IEC 61511 is a process sector implementation of IEC 61508 and part 1 has been released in 2003. The standard comprises three parts and includes extensive guidance on the determination of target safety integrity levels that are to be set by the process design team at the start of the design phase of a protection system.

IEC 61511: Functional Safety: Safety Instrumented Systems for the Process Industry Sector

Part 1: Framework, definitions, system, hardware and software requirements

Part 2: Guidelines in the application of Part 1

Part 3: Guidance for the determination of safety integrity levels

IEC 61511 is directed at the end user who has the task of designing and operating an SIS in a hazardous plant. It follows the requirements of IEC 61508 but modifies them to suit the practical situation in a process plant. It does not cover design and manufacture of products for use in safety, as these remain covered by IEC 61508.

Once IEC 61511 is released the process industries will be able to use it for end user applications whilst devices such as safety certified PLCs will be built in compliance with IEC 61508. IEC 61511 is expected to adopted in the USA and in the EU as the standard for acceptable safety practices in the process industries. ISA S84 will then be superseded.

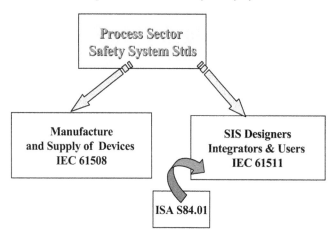

Figure 1.12
Relationship of present and future standards

This diagram shows how S84.01 is the precursor of a process industry sector version of IEC 61508. It came out before the IEC standard but was designed to be compatible with it. Eventually a new standard, IEC 61511, will fulfill the role and S84.01 will possibly be superseded, for the present S84.01 is a very useful and practical standard with a lot of engineering details clearly spelt out. Draft copies of parts of IEC 61511 are incorporating many of the good features set out in ISA S84.01 whilst at the same time aligning its requirements with IEC 61508.

1.10 Equipment under control

The term EUC or equipment under control is widely used in the IEC standard and has become accepted as the basis for describing the process or machinery for which a protection system may be required. The following diagram, Figure 1.13, based on a diagram published in the HSE book 'Out of Control' illustrates what is meant by the term 'equipment under control', abbreviated: EUC.

Scope of Equipment Under Control

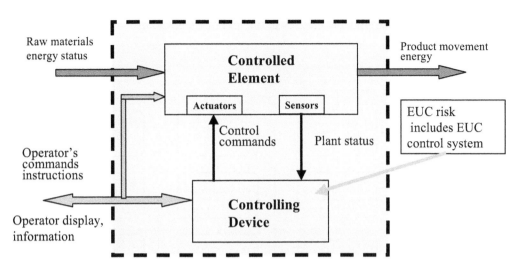

Figure 1.13
EUC

The definition of equipment under control given in the IEC standards is:

'*Equipment, machinery, apparatus or plant used for manufacturing, process, transportation, medical or other activities.*' This includes the EUC control system and the human activities associated with operating the EUC.

This terminology is significant because it makes it clear that the risks we have to consider include those arising from a failure of the control system and any human operating errors.

1.11 The safety life cycle model and its phases (SLC phases)

Introducing the safety life cycle

The foundation for all procedural guidelines in *Safety Instrumented Systems is the Safety Life Cycle (SLC).*

The safety life cycle model is a useful tool in the development of safety related control systems. In concept it represents the interconnected stages from conception through specification, manufacture, installation, commissioning, operation, maintenance, modification and eventual de-commissioning of the plant.

It is visualized by a flow chart diagram showing the procedures suggested for the management of the safety functions at each stage of the life cycle.

1.11.1 Basic SLC

There are a number of versions of the SLC and there is no reason why a particular design team should not draw its own variations. However the standards we have been looking at have drawn up their versions and have laid out their detailed requirements around the framework provided by the SLC.

1.11.2 ISA SLC

Notice how the activities outside of the ISA scope are shown in fainter outlines. See also references to applicable clauses in the text of the standard.

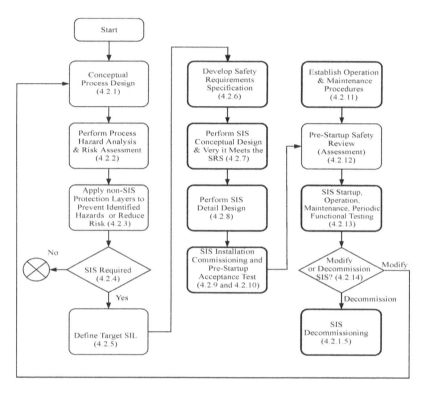

Figure 1.14
ISA SLC

1.11.3 IEC SLC versions

Finally we need to look at the IEC version as this is the most general version and forms the essential core of the IEC standard.

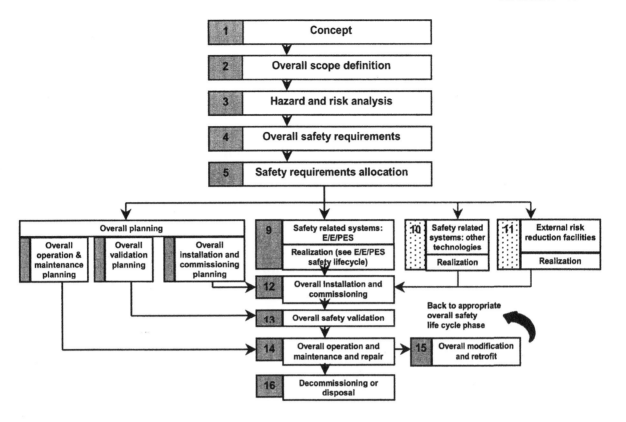

Figure 1.15
IEC SLC version

The IEC SLC indicates the same basic model that we have been considering but adds very specific detail phases as numbered boxes. Each box is a reference to a detailed set of clauses defining the requirements of the standard for that activity. The boxes are easy to follow because they are defined in terms of:

- Scope
- Objectives
- Requirements
- Inputs from previous boxes·
- Outputs to next boxes

Using the SLC assists participants in a safety project to navigate through the procedures needed for the systematic approach we saw earlier

Note the stages of the IEC model. The first 4 phases are concerned with design, then the 'realization' phase is reached. This term describes in very general terms the job of actually building the safety system and implementing any software that it contains.

Once the SIS has been built, the life cycle activities move on to 'installation, commissioning, and validation'. Finally we get to use the safety system for real duties and arrive at the operating and maintenance phase.

In the 'Out of Control' book the HSE provides a commentary on the method of working with the safety life cycle. Like any project model the stages are basically in sequence 'the deliverables of one stage provide the inputs to the next'. However, unlike a project plan the safety life cycle must be regarded as a set of interconnected activities rather than a simple top down design method. It is intended that iteration loops may be carried out at any stage of work; it does not require the completion of one activity before starting another: i.e., 'a concurrent design approach can be used'.

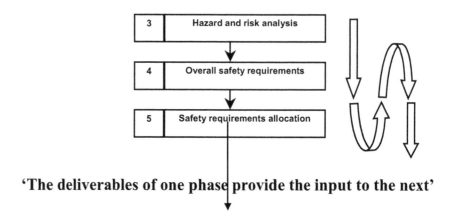

'The deliverables of one phase provide the input to the next'

Figure 1.16
Safety life cycle progression

This shows the idea of a continual iteration between life cycle activities and the verification/assessment task. This is to maintain vigilance that a new activity is always compatible with what has gone before. We might add that this presents a potential nightmare for a project manager!

Large sections of IEC 61508 are concerned with the details of the realization phase and there are whole life cycle models for the activities contained within this stage. Some sections of the IEC standard are dedicated to these specialized tasks. Bear in mind that some of the deeper parts of this standard will be applicable to manufacturers of certified safety PLCs and their associated software packages. A process engineering project would not be expected to dive into such depths.

1.12   Implications of IEC 61508 for control systems

1.12.1   Some Implications of IEC 61508 for control systems

1. This standard is the first international standard that sets out a complete management procedure and design requirements for overall safety control systems. Hence it opens up the way for conformance to be enforced by legislation.

2. Control systems and PLCs serving in safety related applications may be required in the future to be in conformance with the requirements laid down in IEC 61508. Conformance may be required by regulatory authorities before licenses are issued.

3. All forms of control systems with any potential safety implications could be subject to evaluation or audit in terms of IEC 61508.

4. Design and hardware/software engineering of any safety related control system is to be evaluated and matched to required SILs.

5. Integrates responsibility for delivering safety across engineering disciplines, e.g. process engineer, instrument engineer, software engineer, maintenance manager and

maintenance technician are all required to work to the same standard procedures and share all documentation.

6. Software engineering procedures and software quality assurance are mandatory requirements for a PES in safety applications. The standard provides the basis for certification of software packages by authorities such as TUV.

7. Industry specific standards will be derived from guidelines set down in IEC 61508. (Hence all control system safety related applications in any industry may in future be subjected to similar safety life cycle design requirements).

8. Responsibilities of users and vendors are clearly defined:

- The user must define his requirements in terms of functional safety (via the SRS);
- The vendor must show how his solution meets the requirements in terms of the user's specific requirements (compliance with SRS and SIL). It is not sufficient to supply a general purpose ESD logic solver for any application;
- The user's responsibilities for operation, maintenance and change control are defined as part of the conformance.

1.12.2 Potential problems using IEC 61508

W S Black, an IEC working group member, has commented in the *IEE journal*, Feb 2000 on the potential problems some users may face in using the new standard. Some of his points are listed here:

- Deviates from some industry practices
- Sector standards needed to align existing practices e.g. API 41C
- Unfamiliar terminology for USA etc
- Does not match with existing procedures at the start and end of a project
- Project and technical management procedures may need to be redefined to cover key tasks.

1.13 Summary

- The overall design of a safety instrumented system requires that the project participants have a broad knowledge of the hazards and risks as well as the intended protection measures.
- Great care is required in the initial specification stages.
- Successful implementation of a safety system depends on quality assurance in the design process and on good management of all aspects of the project throughout its life cycle.
- The safety life cycle provides the framework for the design and management process.
- New standards describe the procedural and design requirements at each stage of the project life cycle.

1.14 Safety life cycle descriptions

Summary description of safety life cycle phases from HSE's 'Out of Control'

Activities	Objectives,
Concept	To develop a level of understanding of the Equipment under Control (EUC) and its environment (physical, social, political and legislative) sufficient to enable the other safety life cycle activities to be satisfactorily performed.
Scope definition	To determine the boundary of the EUC; To define the scope of the hazard and risk analysis, (eg process hazards, environmental hazards, security considerations such as unauthorized access).
Hazard and risk analysis	To identify the hazards of the EUC and its control system (in all modes of operation) and for all reasonably foreseeable circumstances including fault conditions and misuse; To identify the event sequences leading to these hazards; To determine the EUC risk associated with the identified hazards.
Overall safety requirements specification	To develop the overall safety requirements specification (in terms safety functions requirements and safety integrity requirements) for all safety-related systems and external risk reduction facilities to achieve functional safety.
Safety requirements allocation	To allocate the target safety requirements contained in the overall safety requirements specification (both safety functions requirements and safety integrity requirements) to the designated safety-related systems (SRSs), and external risk reduction facilities.
Overall operation and maintenance planning	To develop an overall operation and maintenance plan to ensure that the functional safety of safety-related systems and external risk reduction facilities are maintained during operation and maintenance.
Planning for overall validation	To develop the overall safety validation plan to enable the validation of the total combination of safety-related systems and external risk reduction facilities to take place.
Planning for overall installation and commissioning	To develop the overall installation plan and the overall commissioning plan so that the safety-related systems and external risk reduction facilities are installed and commissioned in a controlled manner to ensure that the required functional safety is achieved.
Realization of safety-related control systems	To create safety-related control systems (SRCS) conforming to the safety requirements specification (safety functions requirements specification and safety integrity requirements specification).
Realization of external risk reduction facilities	To create external risk reduction facilities to meet the safely functions requirements and safety integrity requirements specified for such facilities.
Overall installation and commissioning	To install and commission the total combination of safety-related systems and external risk reduction facilities
Overall safety	To validate that the total combination of safety-related systems and external risk reduction facilities meet, in all respects, the overall safety requirements

Table 1.2
Safety life cycle phases

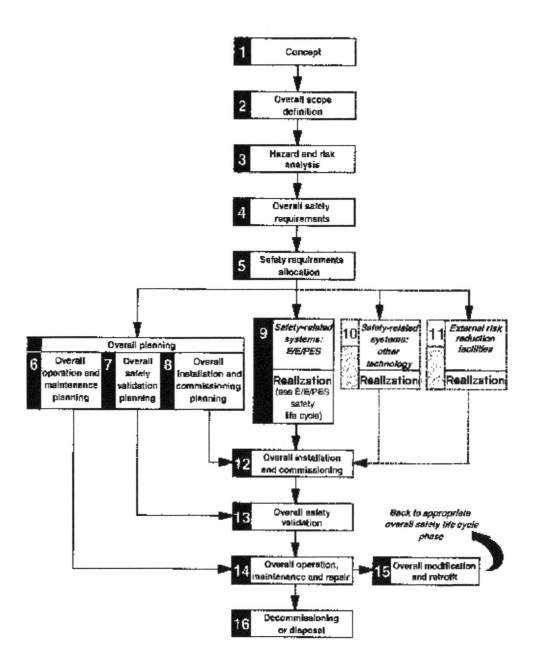

Figure 1.17
IEC overall safety life cycle diagram

1.14.1 Overview of the safety life cycle based on Table 1 of IEC 61508 part 1

Notes:

This table is based on the IEC Table but some phrases have been changed for emphasis; please refer to the IEC standard for exact wording.

SLC phase no. and name	Objectives	Scope	Inputs	Outputs
1 Concept	To develop a level of understanding of the EUC and its environment (physical, legislative etc) sufficient to enable the other safety life cycle activities to be satisfactorily carried out	EUC and its environment (physical, legislative etc)	All relevant information	Information acquired against a checklist given in the std
2 Overall scope definitions	Define the boundaries of the EUC and the EUC control system. To specify the scope of the hazard and risk analysis (e.g. process hazards, environmental hazards, etc)	EUC and its environment	Information acquired in step 1	Information acquired against the phase 2 checklist
3 Hazard and risk analysis	To determine the hazards and hazardous events of the EUC and the EUC control system (in all modes of operation), for all reasonably foreseeable circumstances including fault conditions and misuse. To determine the event sequences leading to the hazardous events determined.	For the preliminary hazard and risk analysis, the scope will comprise the EUC, the EUC control system and human factors. Further h & r analysis may be needed later as the design develops.	Information acquired in step 2	Description of and information relating to the hazard and risk analysis.
4 Overall safety requirements	To develop the specification for the overall safety requirements, in terms of the safety functions requirements and safety integrity requirements.	EUC, the EUC control system and human factors.	Description of, and information relating to, the hazard and risk analysis	Specification for the overall safety requirements in terms of the functions and safety integrity. Includes SIS, non-SIS and external risk reduction measures.
5 Safety requirements' allocation	To allocate a safety function to SIS, non-SIS and external risk reduction measures. To allocate safety integrity, level to each safety function.	EUC, the EUC control system and human factors.	Specification from stage 4.	Allocation decisions for SIS, Non SIS and external measures. Expansion of the SRS for the SIS.
6 Overall operation and maintenance planning	To develop a plan for operating and maintaining the E/E/PE safety-related systems	EUC, the EUC control system and human factors; E/E/PES safety-related systems	As above i.e. Safety requirements Spec.	A plan for operating and maintaining the E/E/PE safety-related systems (SIS)
7 Overall safety validation planning	To develop a safety validation plan for the SIS	As for stage 6	Safety requirements Spec	A plan to facilitate the validation of the SIS

Table 1.3
IEC 61508: safety life cycle

SLC phase no. and name	Objectives	Scope	Inputs	Outputs
8 Overall installation and commissioning	To develop an installation and commissioning plan to ensure the required copy	As for stage 6	Safety requirements Spec	A plan for the installation and commissioning of the SIS.
9 E/E/PE safety related systems realization	To create non-SIS safety systems conforming to the relevant SRS (outside of scope of IEC 61508)	Other technology related systems	Other technology safety requirements spec	Confirmation that each other technology safety-related systems meet the safety requirements for that system.
10 Other technology related systems realization	To create non-SIS safety systems conforming to the relevant SRS (outside of scope of IEC 61508)	Other technology related systems	Other technology safety requirements spec	Confirmation that each other technology safety-related systems meet the safety requirements for that system.
11 External risk reduction factors	To create external risk reduction facilities to meet the relevant SRS (outside of scope of IEC 61508)		External risk, reduction facilities safety requirements specification (outside the scope and not considered further in this standard)	Confirmation that each external risk reduction facility meets the safety requirements for that facility
12 Overall installation and commissioning	To install the E/E/PE safety-related systems; To commission the E/E/PE safety-related systems	E/E/PES safety-related systems	Plans from stage 10	Fully installed SIS

Fully commissioned SIS |
13 Overall safety validation	To validate that the E/E/PE safety-related systems meet the specification for the overall safety requirements	E/E/PES safety-related systems	Plan from stage 9	Confirmation that the SIS meet the safety requirements spec
14 Overall operation, maintenance and repair	To operate, maintain and repair the E/E/PE safety-related systems in order that the required functional safety is maintained	EUC and the EUC control system; E/E/PE safety-related systems.	Requirement for the modification or retrofit under the procedures for the management of functional safety	Continuing achievement of the required functional safety for the SIS; Chronological records of operation repair and maintenance
15 Overall modifications and retrofit	To ensure that the functional safety for the E/E/PE safety-related systems is appropriate, both during and after the modification and retrofit phase has taken place	EUC and the EUC control system; E/E/PE safety-related systems	Request for modification or retrofit under the procedures for management of functional safety	Achievement of the required functional safety for the SIS, both during and after the modification and retrofit phase has taken place; chronological records.

Table 1.3 (cont.)
IEC 61508: safety life cycle

SLC phase no. and name	Objectives	Scope	Inputs	Outputs
16 Decommissioning or disposal	To ensure that the functional safety for the E/E/PE safety-related systems is appropriate in the circumstances during and after the activities of decommissioning or disposing of the EUC.	EUC and the EUC control system; E/E/PE safety-related systems	Request for decommissioning or disposal under the procedures for management of functional safety	Achievement of the required functional safety for the SIS both during and after the decommissioning or disposal activities; Chronological records of activities

Table 1.3 (cont.)
IEC 61508: safety life cycle

1.15 Some websites for safety systems information

Subject	Website www.	Comment
TUV Services in Functional Safety	Tuv-global.com/sersfsafety.htm	Find list of PES certifications Papers on certification etc Detalis of TUV Functional Safety Expert
UK Health and Safety Exec.	hse.gov.uk/sources/index	Range of safety related items + leaflets. UK information sources
Power and Control Newsletter	hse.gov.uk/dst/sctdir.htm	HSE safety specialists provide informative newsletters.
UK Defense Standards	dstan.m.od.uk	Free standards… hazops
IEC	iec.ch/home	Bookstore for IEC 61508
Inst Society of America	isa.org	Bookstore for ISA S84.01
Conformity assessment (CASS)	Siraservices.com	site for conformity assessment training and services
Honeywell SMS	honeywell.co.za/products/ sol_hsms	Safety Management Systems
Hima-Sella UK	hima-sella.co.uk	Hima Range and Applications
Oil and Gas ESD	oil and gas.org	Safety code of practice
Factory Mutual	fm global.com/education_ resources/online_catalog/prssu	Process safety
AIChem Eng …process safety	aiche.org/ccps	
Paul Gruhn	nonvirtual.com/lmeng	US Consultant/Training
Exida	exida.com	US consulting/Engineering Guides
Tony Simmons	tony-s.co.uk	UK Consultant: Safety software
SIS-Tech solutions	sis-tech.com	Consultant. Safety instruments
Siemens	ad.siemens.de/safety	Practices and products
Siemens Moore	moore-solutions.com	Quadlogic and applications
Triconex	tricone xeurope.com	TMR theory and products
UK Directorate of Science	open.gov.uk/hse/dst/sctdir	Software development schemes
Safety Systems R&D	era.co.uk	UK Electrical Research Ass'n see report on offshore safety conference.
Pilz	Pilz.com	Machinery safety systems
Sick	Sick.de	Machinery safety systems

1.16   Bibliography and sources of information

This bibliography contains a list of sources of information relating to safety-instrumented systems or associated activities such as hazard studies.

References used in preparing this book.

Ref No	Title/subject	Origin
1	*Safety Shutdown Systems: Design, Analysis and Justification,* ISBN 1-55617-665-1	Paul Gruhn P.E. and Harry Cheddie P.E., 1998 ISA, PO Box 12277, Research Triangle Park NC 27709, USA www.isa.org
2	*Out of Control: Why control systems go wrong and how to prevent failure.* ISBN 0 7176 0847 6	UK Health and Safety Executive. HSE Books. www.hse.gov.uk
3	*Tolerable Risk Guidelines*	Edward M Marzal: Principal Engineer, Exida .com
4	*Five Past Midnight in Bhopal.* ISBN 0-7432-2034-X	D Lapierre and J Moro. Scribner UK. 2002 (Simon and Schuster UK).
5	*HAZOP and HAZAN* by Trevor Kletz 4th edition. 1999	I Chem. Eng Rugby, UK
6	The design of new chemical plants using hazard analysis. By S B Gibson, 1975	I Chem.E Symposium series no 47. I Chem. Eng Rugby, UK
7	Guidelines on a Major Accident Prevention Policy and Safety Management System, as Required by Council Directive 96/82/EC (Seveso II) ISBN 92-828-4664-4, N. Mitchison, S. Porter (Eds)	European Commission – Major Accident Hazards Bureau. *It is available as a Free download from* Luxembourg: Office for Official Publications of the European Communities, 1998. *Website: www.mahbsrv.jrc.it*
8	IEC 61882: Hazard and Operability Studies (HAZOP studies) – Application Guide. 1st edition 2001-05	International Electro-Technical Commission, Geneva, Switzerland. Download/purchase from: www.iec.ch
9	Hazard and Operability Study Manual. AECI Engineering Process Safety	D Rademeyer Ishecon SHE Consultants Ltd, PO Box 320 Modderfontein, 1645, South Africa.
10	*HAZOP Guide to Best Practice*: by Frank Crawley, Malcom Preston and Brian Tyler. (ISBN0-85295-427-1) Published in 2000 and reprinted 2002.	Published by: European Process Safety Centre, Inst of Chemical Engineers, 165–189 Railway Terrace , Rugby, CV21 3HQ, UK www.icheme.org.uk

11	Alarm systems: A guide to design, management and procurement. EEMUA Publication No 191. 1999	Engineering Equipment and Materials Users Association. UK.
12	IEC 61508 Functional safety of E/E/PES systems. Parts 1 to 7	International Electro-Technical Commission, Geneva, Switzerland. www.iec.ch
13	IEC 61511 Safety instrumented systems for the process industry sector. Parts 1 and 3. 2002.	IEC
14	ANSI/ISA –S84.01 Application of safety instrumented systems for the process industries	ISA.Org
15	DEF 00-55 Hazop studies on systems containing programmable electronics	UK Defence dept. Free download: www.dstan.mod

1.16.1   Suggested books

Dr David J Smith: *Reliability Maintainability and Risk,* 6th edition 2001, Butterworth Heinemann

Trevor Kletz: *HAZOP and HAZAN – Identifying and Assessing Process Hazards*, IChemE, 1999

Felix Redmill, Morris Chudleigh and James Catmur: *System Safety – HAZOP and Software HAZOP*, John Wiley and Sons, 1999

William M Goble: *Control Systems Safety Evaluation and Reliability*, 2nd edition 1998, ISA

Trevor Kletz, Paul Chung, Eamon Broomfield and Chaim Shen-Orr: *Computer Control and Human Error,* I.ChemE, 1995

E.Knowlton: *An introduction to Hazard and Operability Studies – the Guide Word Approach.* Chemetics International, Vancouver, BC, Canada, 1992

1.16.2   Publications

Center for Chemical Process Safety of the American Institute of Chemical Engineers:

Guidelines for Safe Automation of Chemical Processes, AIChE New York 1999, ISBN 0-8169-0554-1

Guidelines for hazard evaluation procedures. ISBN 0-8169-0491-X

Institution of Electrical Engineers

Safety, Competence and Commitment – Competency Guidelines for Safety-related System Practitioners, 1999

Engineering Equipment and Materials Users' Association

Alarm Systems – A guide to design, management and procurement.
EEMUA Publication No. 191, 1999

CASS Limited

The CASS Assessor Guide – The CASS Assessor Competency Scheme,
CASS Ltd 1999

The CASS Guide – Guide to Functional Safety Capability Assessment (FSCA),
CASS Ltd 1999

Simmons Associates

Technical briefs on safety systems. Short descriptions of key issues in safety systems.
Available as free downloads from: Simmons Associates: www.tony-s.co.uk

1.16.3   Reports

Health and Safety Commission

The use of computers in safety-critical applications – final report of the study group on
the safety of operational computer systems, HSC, 1998

Health and Safety Executive

The explosion and fire at the Texaco Refinery, Milford Haven 24, July 1994, 1997

Health and Safety Executive

The use of commercial off-the-shelf (COTS) software in safety-related applications, HSE
Contract Research Report No. 80/1995

1.17   Guidelines on sector standards

Process industry

Reference: IEC 61511
Date: 11th June 1999
Title: Functional safety instrumented systems for the process industry sector
Description: This standard is an adaptation of IEC 61508 for the process industry and
provides details on a general framework, definitions and system software and hardware
requirements.

Part 1: Framework, definitions, system, hardware and software requirements
Part 2: Guidelines in the application of IEC 61511-1
Part 3: Guidelines in the application of Hazard & Risk Analysis

Draft copies of this standard have been in circulation amongst contributing parties but
publication of the approved version is not expected to be complete until later in 2003.
Parts 1 and 2 have been available for purchase from IEC from February 2003. This
standard will be of great value for practical application in the process industries. It
incorporates substantial sections of guide material previously published in ISA S84.01
and is expected to replace ISA S84.01.

Oil and gas industries

Reference: UK Offshore Operators Association
Date: December 1995
Title: Guidelines for Instrument-based Protective Systems
Description: The guidelines have been prepared to provide guidance on good practice for the design, operation, maintenance and modification of instrument-based protective systems on oil and gas installations. The guidelines advocate and translate a risk-based approach to the specification and design of protective instrumentation

Reference: American Petroleum Institute, API 41C 4th edition
Date: (sixth edition)
Title: Recommended Practice for Analysis, Design, Installation and Testing of Basic Surface Safety Systems

Reference: ISO Standard 10418
Date: 1993
Title: Offshore Production Platform – Analysis, Design, Installation and Testing of Basic Surface Safety Systems
Identical content with API 41C 4th edition. Revised version being developed incorporating instrument protection systems to be implemented according to IEC 61508.

Machinery sector

Reference: EN 954 Parts 1 and 2 Draft
Date: March 1997
Title: Safety of Machinery – Safety Related Parts of Control Systems
Description: Parts of machinery control systems are frequently assigned to perform safety functions. Part 1 of this standard provides safety requirements and guidance on the general principles of safety related parts of control systems. Part 2 specifies the validation process including both analysis and testing for the safety functions and categories for the safety-related control systems.

Reference: EN 1050
Date: November 1996
Title: Safety of Machinery – Principles of Risk Assessment
Description: The standard establishes general principles for risk assessment, and gives guidance on the information required to allow risk assessment to be carried out. The purpose of the standard is to provide advice for decisions to be made on the safety of machinery.

Reference: EN 61496 parts 1, 2 and 3.
Dates: 1997–2001
Titles: Safety of machinery – Electro sensitive protective equipment
　Part 1: General requirements and tests.
　Part 2: Particular requirements for equipment using active opto-electronic protective devices (AOPDs)
　Part 3: Particular requirements for Active Opto-electronic Protective Devices responsive to Diffuse Reflection (AOPDDR)

Railway industry

Reference: Cenelec prEN 50126
Date: 27/11/95
Title: Railway Applications – The Specification and Demonstration of Dependability, Reliability, Availability, Maintainability and Safety (RAMS)
Description: This standard is intended to provide railway authorities and the railway support industry throughout the European Community with a process which will enable the implementation of a consistent approach to the management of RAMS.

Reference: Cenelec prEN 50128
Date: 15/12/95
Title: Railway Applications – Software for Railway Control and Protection Systems
Description: The standard specifies procedures and technical requirements for the development of programmable electronic systems for use in railway control and protection applications. The key concept of the standard is the assignment of levels of integrity to software. Techniques and measures for 5 levels of software integrity are detailed.

Reference: Cenelec ENV 50129
Date: May 1998
Title: Railway Applications – Safety Related Electronic Systems for Signaling
Description: ENV 50129 has been produced as a European standardization document defining requirements for the acceptance and approval of safety related electronic systems in the railway signaling field. The requirements for safety related hardware and for the overall system are defined in this standard. It is primarily intended to apply to 'fail-safe' and 'high integrity' systems such as main line signaling.

Medical industry

Reference: IEC 60601-1-4 (2000-04) Consolidated Edition
Date: April 2000
Title: Medical electrical equipment – Part 1-4: General requirements for safety – Collateral Standard: Programmable electrical medical systems
Description: Specifies requirements for the process by which a programmable electrical medical system is designed. Serves as the basis of requirements of particular standards, including serving as a guide to safety requirements for the purpose of reducing and managing risk. This standard covers requirement specification, architecture, detailed design and implementation software development, modification, verification and validation, marking and accompanying documents

Defense standards

These standards are generally available as free downloads from the D Stan website. www.dstan.mod.uk
Reference: 00-54 Pt 1 & 2 Issue 1
Date: 19 March 1999
Title: Requirements for Safety Related Electronic Hardware in Defense Equipment (Interim Standard)
Description:
Part 1 – Describes the requirements for procedures and technical practices for the acquisition of safety related electronic hardware (SREH).

Part 2 – Contains guidance on the requirements contained in Part 1. This guidance serves two functions; it provides technical background to the requirements and it offers guidance on useful techniques for design assurance.

Reference: 00-55 Pt 1 & 2 Issue 2
Date: 1 August 1997
Title: Requirements for Safety Related Software in Defense Equipment
Description: This defense Standard describes the requirements (Part 1) and guidance (Part 2) for procedures and technical practices for the development of safety related software. These procedures and practices are applicable to all MOD Authorities involved in procurement through specification, design, development and certification phases of SRS generation, maintenance and modification.

Reference: 00-56/Issue 2 Parts 1 and 2
Date: 13th December 1996
Title: Safety Management Requirements for Defense Systems
Description: The standard defines the safety program requirements for defense systems. The purpose of part 1 is to define the safety program management procedures, the analysis techniques and the safety verification activities that are applicable during the project life cycle. Part 2 of the standard provides information and guidance to help to implement the requirements of the standard effectively.

Reference: 00-58/Issue 2 Parts 1 and 2
Date: 26th July 1996
Title: Hazop Studies on Systems Containing Programmable Electronics
Description: Part 1 of the standard introduces requirements for processes and practices for hazard and operability studies (HAZOP studies). Part 2 contains guidance on the requirements in order to aid conformance, and also provide procedural background.

Nuclear power

Reference: IEC 61513
Date: March 2001
Title: Nuclear power plants – Instrumentation and control for systems important to safety – General requirements for systems
Description: Provides requirements and recommendations for the total I&C system architecture which may contain either of the following technologies used in I&C systems important to safety: conventional hardwired equipment, computer-based equipment or a combination of both types of equipment.
Note: This standard interprets the general requirements of IEC 61508 parts 1 and 2 for the nuclear industry. It follows similar safety life cycle principles but has deviations from IEC 61508 that are specific to the nuclear power industry.

Furnace safety

prEN 50156 – Electrical Equipment for Furnaces
This draft European standard in preparation will be specific to furnaces and their ancillary equipment. It will cover a number of hardware related items, most of which have been derived from a German standard DIN VDE 0116. Functional safety will be covered in clause 10, which details the safety requirements for electrical systems. This clause will be based on the principles of IEC 61508 and a similar safety life cycle model will be used.

2

Hazards and risk reduction

2.1   Introduction

In this chapter we are going to build up an understanding of hazards and risk and look at some methods available for identifying and quantifying them in any particular situation. In the next chapter we will look more closely at the procedures needed to conduct a hazard study.

As we go along we are going to examine the roles of control systems and safety systems. We need to look at the essential differences between control and safety functions. The feasibility of using instrumentation to solve a given safety problem is one of the reasons why a control engineer or instrument engineer is an essential member of a detailed hazard study team.

Then we look at the principles of risk reduction and the concept of layers of protection. Obviously safety instrumentation is just one of the layers of protection used to reduce risk.

With an understanding of the role of safety instrumented systems in risk reduction we are able to introduce the concept of 'safety integrity' and 'safety integrity level', (SIL).

Identification of hazards, typical sources and examples

A basic understanding of hazards and the potential for hazardous events is essential for persons involved in hazard studies. In other words: it helps if you know what you are looking for!

This book is not intended to deal with hazards in any depth but it may be helpful to have a simple checklist of hazards and perhaps keep adding to it as ideas develop. Please bear in mind that the subject of hazard studies is a large one and that what we cover in this book can only serve as an introduction, sufficient perhaps for you to understand their function. You should at least know what is involved in setting up a hazard study for a process or machine.

Firstly recall the definition of a hazard:

    '*An inherent physical or chemical characteristic that has the potential for causing harm to people, property, or the environment.*'

2.2 Consider hazards under some main subjects:

2.2.1 General physical

- Moving objects.... falling brick, 10-ton truck, meteorite, asteroid?
- Collisions
- Collapsing structures
- Floods

2.2.2 Mechanical plant

- Pressure releasing effects: bursting vessels, jets of gas or liquid, cutting of pipes, thrust forces, noise, shock waves
- Vibration
- Rotating machines, eccentric forces
- Abrasion, grinding, cutting
- Machinery control failure. Cutters, presses, trapping of clothes, conveyor flaps, injection molding etc
- Welding torches, gases and arcs

2.2.3 Materials

- Wrong identities, wrong sizes
- Failures due to stress, corrosion, cracking, chemical attack, fatigue

2.2.4 Electrical

- Flashovers and burns
- Electrocution, power switched on during maintenance
- Fires, meltdowns
- Explosion of switchgear
- Power dips
- Wrong connections, loose connections

2.2.5 Chemical and petroleum

- Explosion
- Fire
- Toxic materials release
- Mechanical failures
- Wrong chemistry: creating the wrong substance through malfunction of the process

2.2.6 Food processing

- Process errors, failure to sterilize
- Contamination of product

2.2.7 Bio-medical/pharmaceuticals

- Toxic release of virus
- Side effects of drugs

- Infection
- Sharps injuries
- Misuse of drugs
- Packaging and labeling errors (wrong drug, wrong gas)
- Process errors, wrong blend, contamination of product
- Sterilization failures

2.2.8   Nuclear power

- Radiation, contamination
- Leakage of toxic materials
- Loss of control
- Excess energy, steam, pressure

2.2.9   Domestic

- Heaters
- Cookers
- Staircase
- Swimming pool
- Criminals

2.2.10   Industries where functional safety systems are common

Our focus is on manufacturing and process industries but hazard studies can be applied to a wide range of circumstances. For our subject of safety instrumented systems we need to look more closely at the industries where active or functional safety systems are commonly found. These are:

- Chemicals manufacturing
- Petroleum refining and offshore platforms, unit protection, platform ESDs, platform isolation ESDs
- Natural gas distribution, compressor stations
- Marine, ship propulsion, cargo protection
- Power generation (conventional and nuclear)
- Boilers and furnaces
- Mining and metallurgy processes
- Manufacturing and assembly plants, machinery protection systems
- Railway train control and signaling systems

2.3   Basic hazards of chemical process

Let us take the case of chemical plant hazards and see how they are related to the hazard study methods. These give us an idea of how a combination of conditions must be tested to reveal conditions that could cause an accident.

2.3.1   Some causes of explosions, fire and toxic release

Causes of explosions: In order to create an explosion it is necessary to have together an explosive mixture or material and an ignition source. An explosion inside a container or in building is said to be 'confined' whilst an explosion in the open air is 'unconfined.'

Explosive mixture: Occurs when a gaseous fuel is mixed with an oxidant (Air, oxygen). The proportions of the mixture must lie between the lower explosive limit (LEL) and the upper explosive limit (UEL). Between the LEL and the UEL there is sometimes a region where the mixture is so potent that when ignited it will detonate which means that the flame front will form a shock wave and travel at the speed of sound. Outside of the detonation range the flame front will travel slowly at typically 1 m/s and the explosion is described as a deflagration. Hence the difference between a 'bang' and a 'whoosh!'.

The Table 2.1 gives explosion and detonation limits for some common gases in air.

GAS	EXPLOSIVE		DETONATION	
	Lower (%)	Upper (%)	Lower (%)	Upper (%)
Hydrogen	4	74.2	18.3	59
Acetylene	2.5	80	4.2	50
Methane	5	15	10*	53
Butane	1.18	8.4	2	6.2
Methanol	6.7	36.5	9.5*	64.5
Ether	1.9	36.5	2.8	4.5
Ammonia	16	27	25*	75*

*Only in pure oxygen

Table 2.1
Explosion and detonation limits

Similarly there are explosion boundaries for flammable dusts as shown in the next table.

MATERIAL	MINIMUM CONCENTRATION IN AIR (g/m^3)
Coal	55
Sulphur	35
Polythene	20

Table 2.2
Explosion boundaries for flammable dusts

2.3.2 Logic diagram for an explosion

An explosion of a gas cloud requires the combination of the explosive gas mixture and a source of ignition, unless the mixture is able to detonate on its own. Hence an elementary logic diagram will look like this one in the Figure 2.1.

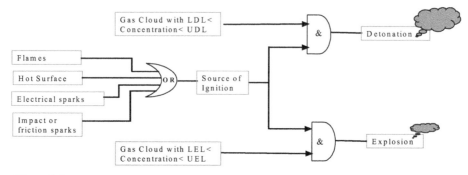

Figure 2.1
Logic diagram for an explosion

Logic diagrams are used as the basis for representing combinations of hazard conditions and faults, they are generally known as 'fault tree diagrams'. We shall see more of them later.

Sources of ignition will include:

- Flames and hot surfaces
- Sparks from electrical or mechanical or chemical causes
- Static electricity
- Compression, impact or friction

Ignition of any particular gas mixture requires:

- An energy level above a specific minimum level
- A temperature above specific minimum

Hence, it becomes possible to prevent electronic instrumentation from becoming a source of ignition in the plant areas if its temperature and stored energy levels can be limited by design, to values below the thresholds for the categories of gases on the plant. We know this from of protection as INTRINSIC SAFETY. This is a form of hazard prevention. It must not be confused with a 'safety instrumented system' but it does feature in the list of risk reduction measures.

2.3.3 Fires: causes and preventative measures

Chemical fires can occur inside process containment but are often more severe when they occur outside of the containment due to leakages or incorrect discharges of flammable materials. Again a source of heat or ignition is required but in the case of fires the combustion is sustained by steady vaporization of the fuel rather than an instantaneous explosion.

Prevention of fires is a form of hazard protection and the chances of avoiding ignition of fuels appear to be greater than is the case with flammable vapor clouds. Hence the elimination of sparking sources through explosion protection of electrical power equipment is a major protection measure. As noted before intrinsically safe instrumentation is another risk reduction measure.

Equipment used to detect and extinguish fires can be classified as active safety systems and thus may fall within the scope of functional safety. If we look again at the definition of functional safety we can see that failure to operate a fire and gas detection system would definitely pose a danger to people and equipment. It begins to look as if the requirements laid down in IEC 61508 are applicable to fire and gas detection systems.

2.3.4 Toxic material release

There are many forms of toxic materials ranging from the drastic (asphyxiates) to the nuisances and including long-term effects of low intensity pollutants. In general safety instrumented systems see service in the prevention of critical and large scale releases whilst the longer-term toxicity hazards are prevented by plant design and operating practices.

2.3.5 Failures of equipment

Many of the applications for SIS are associated with hazards created by operational faults or equipment failures and associated with the threat of explosion, fire or toxic release.

It is therefore useful to be aware of typical failures and operational faults seen in process plants.

Typical failures and operational faults seen in process plants

This list is compiled from the AECI process safety manual.

- Materials of construction
- Equipment failures, list of heat sources, pumps etc
- Instrumentation...table of failure, modes, effects of a failed control loop
- Fails to operate the process within design limits
- Operator failure...tables of error rates and reaction times...absence of operators
- Maintenance related failures
- Modifications to design, temporary arrangements

2.4   Introduction to hazard studies and the IEC model

As we have seen specification errors contribute a large proportion of safety system failures. Recognizing and understanding the safety problem to be solved is the first essential step in avoiding this problem. This in turn requires that we understand the nature of hazards and the contributing factors. Hence the emergence of systematic hazard study methods.

2.4.1   Introduction to hazard studies

The foundation for any safety system application is a thorough understanding of the problem to be solved. The process industry seems to have reached consensus on the use of a top down methodology that was developed about 30 years ago and is generally known as the hazard study method.

The studies begin at the earliest possible stage of a project when the concept or outline of the process is set down and are repeated in increasing detail as the design and building of the plant evolves. By this method, potential hazards are identified at the earliest possible stage in a design when corrective actions are easiest to carry out. Further levels of study, performed as the design detail improves, serve to verify the safety measures taken so far and identify potential problems emerging in the details.

2.4.2   Alignment with the IEC phases

The first 3 levels of process hazard studies align closely with the first three phases of the IEC safety life cycle model.

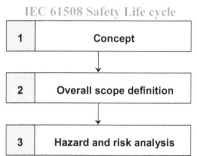

Figure 2.2
IEC 61508 safety life cycle

It is interesting to note how the IEC model has captured the essence of existing well-tried practices and has arrived at a version that is hopefully intelligible and acceptable to a wide variety of industries and companies. Let's now look at the IEC Phases 1, 2 and 3 as extracted from the SLC model in the IEC Standard.

2.4.3   Box 1: Concept

Objective

- Preliminary understanding of the EUC and its environment (process concept and its environmental issues, physical, legislative etc)

Requirements

- Acquire familiarity with EUC and required control functions
- Specify external events to be taken into account
- Determine likely source of hazard
- Information on nature of the hazard
- Information on current applicable safety regulations
- Hazards due to interaction with other equipment/processes

2.4.4   Box 2: Scope definition

Objectives

- Determine the boundary of the EUC and the EUC control system
- Specify the scope of the hazard and risk analysis

Requirements

- Define physical equipment
- Specify external events to be taken into account
- Specify sub systems associated with the hazards
- Define the type of accident initiating events to be considered (e.g. component failures, human error)

2.4.5   Box 3: Hazard and risk analysis

Objectives

- Determine the hazards and hazardous events of the EUC and the EUC control system...all foreseeable circumstances
- Determine the event sequences leading to the hazardous events
- Determine the EUC risks associated with the hazardous events

Requirements

- Hazard and risk analysis based on the scope defined in phase 2
- Try to eliminate the hazards first
- Test all foreseeable circumstances/abnormal modes of operation
- Sequence of events leading to the hazard event + probability
- Consequences to be determined

- Evaluate EUC risk for each hazardous event
- Use either qualitative or quantitative assessment methods

Choice of technique depends on:

- Nature of the hazard
- Industry sector practices
- Legal requirements
- EUC risk
- Availability of data
- Hazard and risk analysis to consider
- Each hazard and events leading to it
- Consequences and likelihood
- Necessary risk reduction
- Measures taken to reduce the hazard or risk
- Assumptions made to be recorded
- Availability of data
- References to key information
- Information and results to be documented (hazard study report)
- Hazard study report to be maintained throughout the life cycle

All of which identifies the IEC hazard and risk analysis requirements as being very much the same as the process hazard analysis level 3 or Hazop study.

2.4.6 Conclusions

Hazard studies

- Allow a progressively deeper level of knowledge of a process or equipment function to be obtained as the design or the knowledge becomes available
- Offer the opportunity to reduce potential hazards and risks at the earliest possible stages of a design
- Form part of the quality management system needed to provide the best possible management of any risks inherent in the process
- Have been incorporated into the IEC safety life cycle to provide a common baseline for all safety activities

Now that we have some idea of the role of hazards, let's look at the practical risk reduction side of the equation.

2.5 Process control versus safety control

At this point we feel it is appropriate to take a brief look at the issue of process control versus safety control.

2.5.1 Historical

Safety functions were originally performed in different hardware from the process control functions. This was a natural feature because all control systems were discrete single function devices. It was not really inconvenient for instrument designs to achieve the separation and extra features needed for the safety shutdown devices.

Only with the advent of DCS and PLC controllers did we have to pay attention to the question of combining safety and control in the same systems. There is a temptation nowadays to combine safety and control functions in the same equipment and it can be acceptable under certain conditions but it should be resisted wherever possible. It also depends on what is meant by 'combine'.

2.5.2   Separation

All standards and guidelines clearly recommend the separation of the control and safety functions. Here are some examples from the various standards and guidelines to back up this statement.

This diagram is taken from the UK Health and Safety Executive's publication: *'Programmable Electronic Systems in Safety Related Applications'*.

Basic reasons for separation are clear from the next two figures.

Process Control versus Safety Control

Separation of safety controls from process controls

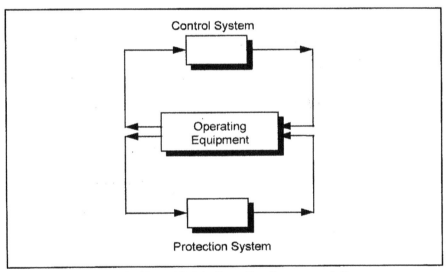

U.K. Health and Safety Executive Recommendations

Figure 2.3
Process control versus safety control

Quotation from IEEE

'The safety system design shall be such that credible failures in and consequential actions by other systems shall not prevent the safety system from meeting the requirements.'

Quotation from ANSI/ISA S 84.01

'Separation between BPCS and SIS functions reduces the probability that both control and safety functions become unavailable at the same time, or that inadvertent changes affect the safety functionality of the SIS.'

'Therefore, it is generally necessary to provide separation between the BPCS and SIS functions.'

2.5.3   Functional differences

Some basic differences between the two types of control system support the idea that control and safety should be separated.

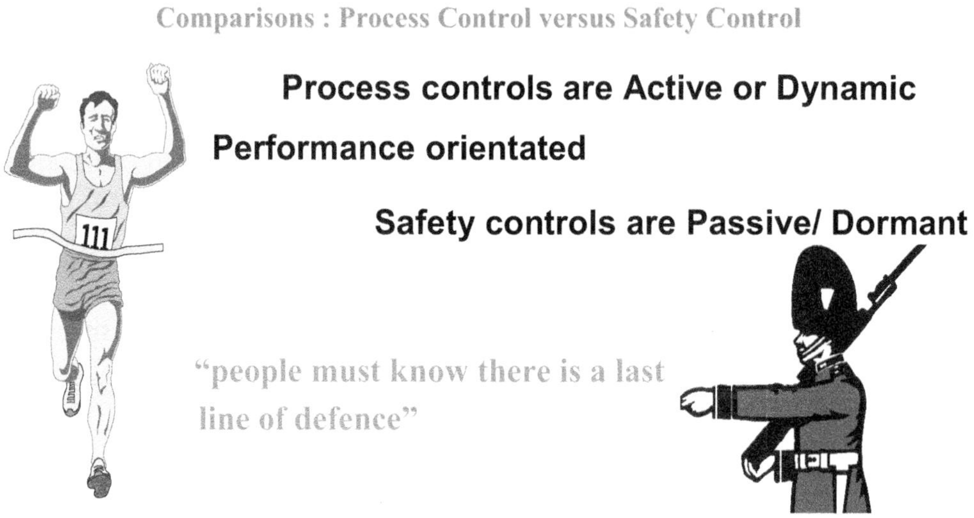

Comparisons : Process Control versus Safety Control

Process controls are Active or Dynamic

Performance orientated

Safety controls are Passive/ Dormant

"people must know there is a last line of defence"

Figure 2.4
The roles of process control and safety control

Process control versus safety control

Process controls act positively to maintain or change process conditions. They are there to help obtain best performance from the process and often are used to push the performance to the limits that can safely be achieved. They are not built with safety in mind and are not dedicated to the task.

Because they are operating at all times they are not expected to have diagnostic routines searching for faults. Generally a fault in process control is not catastrophic although it can be hazardous, hence the need to include the EUC control system in the risk analysis.

Most significantly one of the most valuable assets of a process control system is its flexibility and ease of access for making changes. Operators may need to disable or bypass some portions of the control system at different times. This is exactly what we want to avoid in a safety control system.

Safety controls are passive/dormant

Safety controls are there as policemen and security guards. They need to be kept to a fixed set of rules and their access for changes must be carefully restricted: i.e. they should be incorruptible. And they must be highly reliable and be able to respond instantly when there is trouble!

Comparisons: process control versus safety control

This table summarizes some of the differences between process control and safety control systems.

Feature	Process control	Safety control
Control type	Active, complex, optimizing	Passive, simple, direct acting
Tasks	Many, variable, expanding, experimental	Limited, strictly defined
Modes of control	Auto, manual, supervisory	Auto, no manual interventions, no external command levels
Communications	Open systems Fieldbus/Profibus etc	Limited, specialized Difficult with bus networks
Changes	Easy to make, password protected configurable, parameter changes	Strictly controlled, password protected, verified and documented. Parameter changes strictly controlled
Diagnostics	Limited	Intensive. Proof testing
Redundancy	Used for high availability	Used for high reliability
Documentation	For convenience	Essential for validation of each function
Testing	Nominal loop testing Detailed sequential	Failure modes tested
Legal	Not regulated	Subject to regulation, audit and certification

Table 2.3
Differences between process control and safety control systems

2.5.4   Specials: integrated safety and control systems

Having said so much in favor of separation of control and safety functions there are always going to be situations where the control system also performs a safety role or where an integrated package is needed. An example would be a large turbo compressor set where anti surge control and load control are integrated into one scheme.

Burner management systems are another example where the fuel/air ratio controls provide a safety role and where the start up ignition and shutdown purge sequences protect against possible explosions of unburnt fuel in the furnace.

What if a process controller has a safety function?

The philosophy in the new standards is clear on this issue. If the control system performs a safety-related function it should be considered to be a safety instrumented system. This leaves only two options as shown in the following figures.

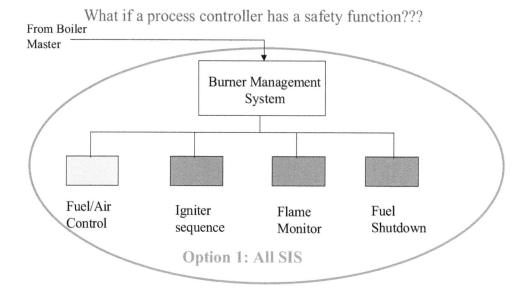

Figure 2.5
All SIS

Option 1, All SIS

or

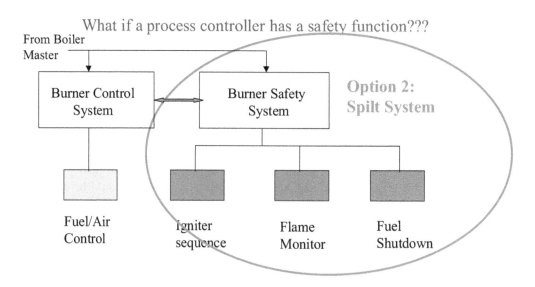

Figure 2.6
Split system

Option 2, split system

In cases where the process control functions are well defined the solution is often to use option 1 and build a single high integrity control system. Internal architectures of the SIS support subdividing of high and low integrity safety functions for economy.

2.6   Simple and complex shutdown sequences, examples

At this point it may be helpful to illustrate some typical features of safety instrumented systems before we move on to risk reduction concepts.

2.6.1   Simple shutdown sequence

A simple shutdown device is shown in the next two diagrams.

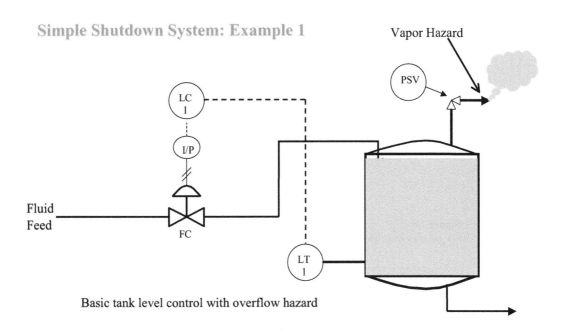

Figure 2.7
Simple shutdown system: hazard

The above diagram shows a flammable liquid being drawn from a process source into a buffer tank from which it is to be pumped onwards to a treatment stage. A typical level control loop is provided in the process control system to maintain level at say 50% full. A hazard will occur if the level control fails for any reason and the tank becomes full. The tank must have a pressure relief valve by law and if the liquid has to escape via this valve it will form a dangerous vapor cloud.

The reasons for loss of level control include:

- Jammed open level valve (dirt or seizure)
- Failed level transmitter: giving a false low level signal
- Control loop left on manual with the valve open
- Leaking control valve with the pump out stage shutdown

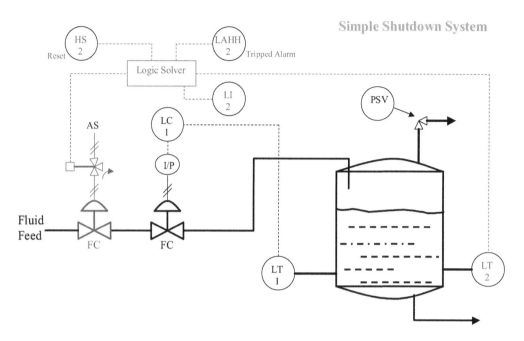

Figure 2.8
Simple shutdown system: solution

A simple instrumented shutdown device would require the features shown in the above diagram: i.e. a level switch set to detect extra high level in the tank causes an automatic shutoff valve to close off all liquid feed to the tank. The shutoff valve remains closed until the defect in the process control system has been rectified.

Once the level has been restored to normal, the operator is allowed to reset the shutoff device and continue operations.

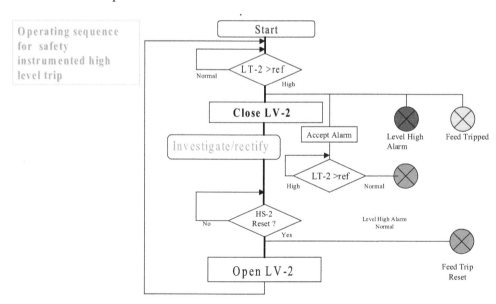

Figure 2.9
Operating sequence for safety instrumented high-level trip

Note how even the simplest of protection systems requires some degree of logic functionality. Hence the logic solver role in the SIS quickly becomes complex.

2.6.2   Complex shutdown sequences

Example of Process Requiring Complex Trip Sequence

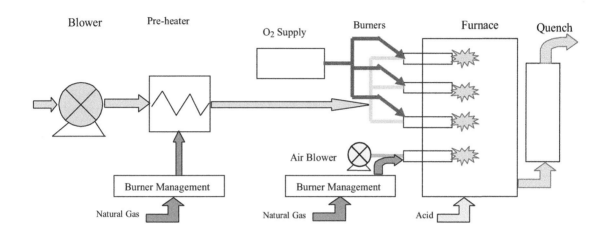

Figure 2.10
Example of process requiring complex trip sequence

This example is based on a section of a large ICI chemical process plant. The Figure 2.10 shows a process gas blower (steam turbine driven) delivering gas via a preheater stage into a set of burners in a processing furnace. Natural gas is used for preheating and for base heating for the furnace. Oxygen is injected into the gas burner stages; also an acid stream is sprayed into the furnace to neutralize the product. The hot gases are then passed via a quench tower before being delivered into the next stage of the process.

The safety systems for this plant will be typical for a large processing plant. There are some common characteristics of the safety systems for such plants and this example illustrates them.

Local process conditions

Each stage of the process has its own inherent hazards and limiting conditions. Hence local process sensors are used to detect process hazards in each stage. Actuated control valves and electrical trip signals are then used to shut down the stage that has the problem.

Machinery protection

In those stages where high performance machinery is used, for example the gas blower, the machine itself will have numerous sensors to protect against the hazards and costs involved when the machine develops a fault.

For example the blower will have detectors for vibration, temperature and displacement of the bearings. If it were electrically driven there would be temperature detectors for the windings of the motor.

Burner management

Where burners are used there is normally a complete burner management control system with its own set of protective instruments and a safety control system.

Stages are interdependent

Each stage in this process is dependent on the correct working of at least one other stage. Hence the safety instrumented system for any one stage is likely to have additional shutdown commands from the status of the other critical stages. The trip logic diagram shown in the Figure 2.11 shows how the shutdown or tripping of one stage leads to the tripping of another. The entire plant will shutdown if the quench tower process trips out. However if one burner trips out in the furnace the logic diagram shows that the plant will continue operating.

Complex Trip Sequence for Process Example

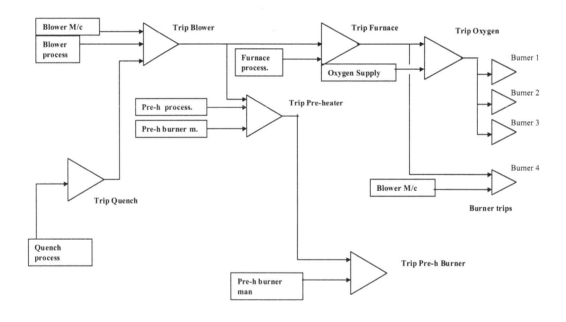

Figure 2.11
Complex trip sequence for process example

All the trip logic is performed in the safety system

It is important to note that all the actions required by the trip logic diagram have been evaluated through hazard studies as being essential for the safety of personnel and plant. Hence the entire system shown is a safety system and remains independent of the basic process control system for the plant. The trip system plays no role in the start up of the plant except that it will prevent the start up of any one stage if it is not safe to do so by virtue of the trip logic. We shall be returning to this example later in the course when we look at design features. What we are trying to show by the examples given is the typical nature of safety instrumented systems as applied to process operations.

2.7   Protection layers

The concept of protection layers applies to the use of a number of safety measures all designed to prevent the accidents that are seen to be possible. Essentially this concept identifies all 'belts and braces' involved in providing protection against a hazardous event or in reducing its consequences.

Layers of Protection

IEC 61511-3 draft

A Protection Layer consists of a grouping of equipment and/or administrative controls that function in concert with other protection layers to control or mitigate process risk.

An independent protection layer (IPL):
- Reduces the identified risk by at least a factor of 10.
- Has high availability (.0.9)
- Designed for a specific event
- Independent of other protection layers
- Dependable and auditable

Figure 2.12
Layers of protection

IEC 61511 has drafted a definition of protection layers for use in risk assessment models. A summary is shown in the above figure. This definition is rigorous because it supports the formalized use of a safety layer in the determination of safety integrity levels for SIS designs.

Layers of Protection

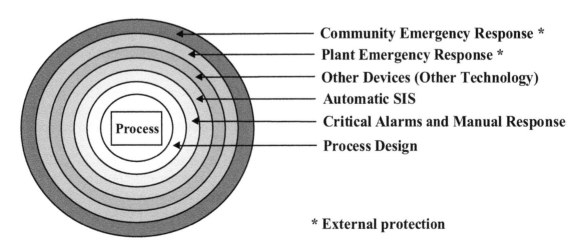

- **Community Emergency Response ***
- **Plant Emergency Response ***
- **Other Devices (Other Technology)**
- **Automatic SIS**
- **Critical Alarms and Manual Response**
- **Process Design**

* External protection

Figure 2.13
Typical collection of protection layers

Protection layers can be divided into two main types: prevention and mitigation as seen in Figure 2.12. We need to look into this concept in some more detail but before we do that it may be of interest to consider a lesson to be learned from the events leading to the disastrous chemical accident at Bhopal in India where somewhere between 2000 and 16 000 people died from toxic gas inhalation after several layers of protection failed at the Union Carbide pesticides plant. These notes are based on the account of the accident described in the book 'Five Past Midnight in Bhopal' by D Lapierre and J Moro and also summarized by Gruhn and Cheddie.

The plant had 3 storage tanks for methyl isocyanate (MIC), an unstable liquid that decomposes into a range of toxic components as its temperature rises above 15° C. Most deadly of these is hydrocyanide acid or cyanide gas, which when inhaled invariably leads to death in a very short time.

The safety systems for the tanks comprised 4 protection layers:

- Each tank was to be operated at no more than 50% capacity to allow room for a solvent to be added in case a chemical reaction started in the tank.
- The tank contents were to be kept below 15° C by means of a refrigerant system circulating Freon through cooling pipes at the tanks. A high temperature alarm was provided on each tank to alert operators to an abnormal temperature rise.
- Should any gases start to emerge from the tanks, they should be absorbed by caustic soda injection as they pass through a decontamination tower.
- Finally if any gases escape the absorber tower a flare at the top of a 34-meter flare stack should burn them off.

Due to a longstanding lack of demand for the pesticides produced by the plant there had been a long period of time when production had been shutdown or kept to a minimum. The plant equipment and operating standards had been allowed to deteriorate. Finally on the night of 2/3 December 1984, the tanks appear to have been contaminated with hot water from a pipe-flushing task. This led to an uncontrollable reaction, which ruptured the tanks, the first of which being 100% full contained 42 tons of MIC. The resulting gas clouds blew across the slum areas adjoining the factory fence and onwards into the city. The death toll is disputed but is claimed by Lapierre and Moro to be between 16 000 and 30 000 with around 500 000 people injured.

How could 4 layers of protection be defeated? The simple answer is that there was a common cause failure that was not factored into the safety calculations and failure to manage the plant according to intended safety and maintenance practices. The individual failures were:

- Tanks were not kept below 50% full as intended. This appears to have been because the nitrogen pressurizing required to operate liquid transfers was so full of leaks that it could not be operated. There was no capacity or awareness of the solvent injection option.
- The refrigeration system had been turned off months earlier including the alarm system because the plant manager did not believe it was necessary to keep the MIC at 5° C. The ambient temperature was 20° C
- The decontamination tower was offline for maintenance and had been so for a week.
- The flare stack was also out of service for maintenance. In one sense it can be argued that the tower and the stack were a combined safety layer and hence were not to be considered as independent layers of protection.

The chemical industry has hopefully learned a lot of hard lessons from the Bhopal disaster but it is informative to read the details and see how familiar are the problems reported from that experience.

2.7.1   Prevention layers

Prevention layers try to stop the hazardous event from occurring.
Examples follow:

Plant design

As far as possible, plants should be designed to be inherently safe. This is the first step in safety and techniques such as the use of low-pressure designs and low inventories are obviously the most desirable route to follow wherever possible.

Process control system

The control system plays an important role in providing a safety function since it tries to keep the machinery or process within safe bounds. As we have noted earlier the control system cannot provide the ultimate safety function since that requires independence from the control system. In fact the accepted practice as we shall see later is to evaluate the risks of any operating process in terms of the EUC working with its control system (EUC Control System).

Alarm systems

Alarm systems have a very close relationship to safety shutdown systems but they do not have the same function as a safety instrumented system. Essentially alarms are provided to draw the attention of operators to a condition that is outside the desired range of conditions for normal operation. Such conditions require some decision or intervention by persons. Where this intervention affects safety, the limitations of human operators have to be allowed for.

Here are some fundamental types of alarm:

- **Process alarms:** These alarms may deal with efficiency of the process or indicate defects in the equipment. This type of alarm is normally incorporated into the plant control system (typically a DCS) and shares the same sensors as the control system.
- **Machinery or equipment alarms:** These alarms assist with detection of problems with equipment and do not directly affect the operation of the process.
- **Safety related alarms:** These alarms are used to alert operators to a condition that may be potentially dangerous or damaging for the plant. Such alarms should normally have a high priority and where they are involved in protecting against maloperation by the control system they should be independent of the devices they are monitoring. These alarms often serve as a pre-alarm to action that will be taken by the SIS if the condition becomes more severe. In many cases this type of alarm is generated by the safety shutdown system itself and this is a good approach because independence from the main control system is assured. There is a disadvantage in that the risk of common cause failure between the SIS and the pre-alarm system must be considered in the reliability analysis of the overall protection system.

- **Shutdown alarm:** This type of alarm tells the operator that an automatic shutdown event has been reached and has been initiated by the SIS. It is basically a monitoring function for the SIS and contributes as a layer of protection because it supports corrective and subsequent actions by the operator.

Issues of alarm systems

There are a number of important issues concerning alarms that we need to touch on in this workshop and we shall return to this subject in one of the later chapters.

Mechanical or non-SIS protection layers

A large amount of protection against hazards can be often be performed by mechanical safety devices such as relief valves or overflow devices. These are independent layers of protection and play an important role in many protection schemes.

Interlocks

Interlocks provide logical constraints within control systems and often provide a safety related function. These functions may be embedded within the basic control system in the form of software or they may be relay or mechanical interlocks directly linked to the equipment. They can therefore be considered as providing a layer of protection and their degree of independence must be evaluated in each particular application – in each application decide if the interlock is part of the control system, an independent safety device or part of the SIS.

Shutdown systems (SIS)

The final protection layer is the safety shutdown system in which automatic and independent action is taken by the shutdown system to protect the personnel and plant equipment against potentially serious harm. The essence of a shutdown system is that it is able to take direct action and does not require a response from an operator.

2.7.2 Mitigation layers

Mitigation layers reduce the consequences after the hazardous event has taken place. They include fire & gas systems, containments and evacuation procedures.

2.7.3 Diversification

Using more than one method of protection is generally the most successful way of reducing risk. The safety standards rate this approach very highly and it is particularly strong where an SIS is backed up with, say, a mechanical system or another SIS working on a completely different parameter.

2.8 Risk reduction and classification

Now we should take a look at the models and explanations given for risk reductions that are set out in the IEC 61508 standard, *Risk Reduction and Safety Integrity as per IEC Annex A1*.

The IEC standard explains that the layers of protection are effectively separate risk reduction functions. The standard presents a general risk reduction model as shown in figure A1.

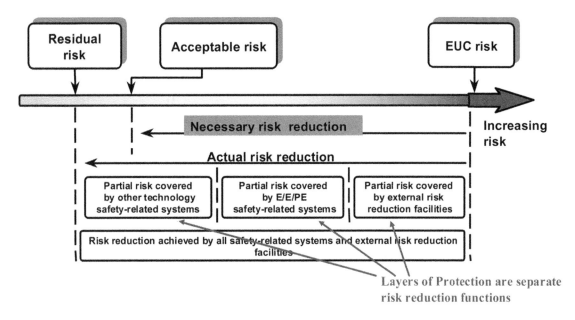

Figure 2.14
Risk reduction: general concepts (from IEC 61508 Part 5)

Risks indicated in Figure A1 from the IEC standard are:

- EUC risk for the specified hazardous event including its control system and the human factors
- Tolerable risk: based on what is considered acceptable to society
- Residual risk: being the risk remaining after all external risk reduction facilities, SIS and any other related safety technology has been taken into account

Risk classification

The problem is that risk doesn't come in convenient units like volts or kilograms. There is no universal scale of risk. Scales for one industry may not suit those in another. Fortunately the method of calculation is generally consistent and it is possible to arrive at a reasonable scale of values for a given industry. As a result IEC have suggested using a system of risk classification that is adaptable for most safety situations. We refer here to: Annex B of IEC 61508 part 5 that provides risk classification tables.

Risk Classification of Accidents: Table B1 of IEC 61508-5

Frequency	Consequences			
	Catastrophic	Critical	Marginal	Negligible
Frequent	I	I	I	II
Probable	I	I	II	III
Occasional	I	II	III	III
Remote	II	III	III	IV
Improbable	III	III	IV	IV
Incredible	IV	IV	IV	IV

Table 2.4
Example of a general risk classification table (based on table B1 of IEC 61508)

The risk classification table is a generalized version that works like this:

- Determine the frequency element of the EUC risk without the addition of any protective features (Fnp)
- Determine the consequence C without the addition of any protective features
- Determine, by use of Table B.1, whether for frequency F np and consequence C a tolerable risk level is achieved

If, through the use of Table B.1, this leads to risk Class I, then further risk reduction is required. Risk class IV or III would be tolerable risks. Risk Class II would require further investigation.

NOTE: Table B.1 is used to check whether or not further risk reduction measures are necessary, since it may be possible to achieve a tolerable risk without the addition of any protective features.

In practice this Table B1 is a generic table for adaptation by different industry sectors. It is intended that any given industry sector should insert appropriate numbers into the fields of the table and hence establish acceptable norms. For example in the next figure some trial values have been inserted.

Frequency	Catastrophic > 1 death	Critical 1 death or injuries	Marginal minor injury	Negligible prod loss
1 per year	I	I	I	II
1 per 5 years	I	I	II	III
1 per 50 years	I	II	III	III
1 per 500 years	II	III	III	IV
1 per 5000 yrs	III	III	IV	IV
1 per 50000 yrs	IV	IV	IV	IV

Table 2.5
Risk classification of accidents: trial values

This type of table is generally known as a 'risk matrix' and many companies have generated their own version for guidance on risk management within their organizations. By writing a scale of numbers into the frequency and consequence scales each hazard can be evaluated and placed into the matrix with a 'risk ranking value'. This results in the typical risk ranking table shown in Table 2.6.

Frequency /yr	Consequences			
	Minor: 1	Significant: 3	Major: 6	Catastrophic: 10
Frequent: 10	6	10	60	100
Probable: 8	8	24	48	80
Possible: 4	4	12	24	40
Unlikely:2	2	9	12	20
Remote: 1	1	3	6	10

10
1
10^{-1}
10^{-2}
10^{-3}
10^{-4}

Unacceptable Region

Transitional Region

Tolerable Region

Table 2.6
Example of a risk ranking matrix

Assuming we have a measure of risk let's try risk reduction from EUC 'risk to tolerable risk'. Using the IEC generalized table (Table 2.4) we can see that to reduce the risk from class 1 to class 3 for a critical consequence risk such as 1 death we would need to reduce the frequency (of the unprotected hazardous event) from 1 per year (Fnp =1) to at least as low as 1 per 500 years (Ft = 0.02). This reduction in frequency or risk is given the term R where :

$$R = Ft/Fnp, \text{ hence in this case } R = 0.02$$

The principle is illustrated in the next figure taken from Figure C1 of Annex C:

Safety Integrity Allocation Example based on IEC Figure C1

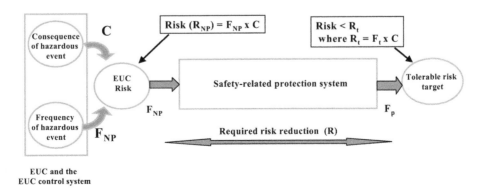

Figure 2.15
Safety integrity allocation example

So the required risk reduction can be defined for each application.

We can go one step further and allocate a measure of risk reduction to each layer of protection that has been identified for a given hazard. The next figure illustrates this allocation where it shows 3 layers of protection, each layer having its own contribution to risk reduction.

Basic risk reduction model

This type of diagram is called a risk reduction model and is a powerful method of showing the details of risk reductions for any application.

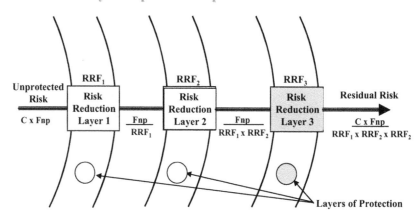

Figure 2.16
Basic risk reduction model

2.9   Risk reduction terms and equations

There is a choice of the terms we can use to define the risk reduction as shown in the next figure.

Risk Reduction Terms and Equations

$$F_t = \text{Tolerable Risk Frequency}$$
$$F_{np} = \text{Unprotected Risk Frequency}$$
$$F_p = \text{Protected Risk Frequency}$$

The Risk Reduction Factor: RRF
$$RRF = F_{np} / F_t$$

Safety Availabilty: $SA\% = (RRF-1) \times 100 / RRF$

Probability of Failure on Demand: PFDavg.
$$PFD_{avg} = 1 / RRF = \Delta R = F_t / F_{np}$$

Figure 2.17
Risks reduction terms and equations

The following notes and the Figure 2.17 should help clarify the use of these terms.

Risk reduction factor. It is often more convenient to talk in terms of risk reduction factor.

We must just be careful to note that: RRF = 1/ •R or RRF = Fnp/Ft

Hence in our example above when the tolerable frequency of a hazardous event is Ft = 0.02/yr and the unprotected hazard frequency is Fnp = 1/yr the RRF = 500

Safety availability. Another term used to express the performance requirements of a risk reduction system is 'safety availability. This is simply the percentage of time that the protection system has to be available to meet the risk reduction target. As you would expect this figure is typically a high percentage

Required safety availability = (RRF − 1) × 100/RRF

For our RRF example of 500 the SA will be: 99.8%

2.9.1 Introducing the average probability of failure on demand...PFDavg

The average probability of failure to perform the design function on demand (for a low demand mode of operation).

It is important to be clear on the concept of probability of failure on demand. The Figure 2.18 illustrates the principle.

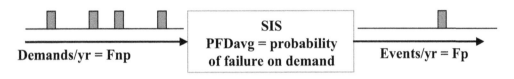

$$Fp = Fnp \times PFDavg$$

Target value for Fp is the tolerable risk frequency Ft

PFDavg is also known as FDT: Fractional Dead Time

Figure 2.18
Introducing PFDavg

The alternative name for PFD is the fractional dead-time, FDT. The meaning is clear: the fraction of time that the safety system is dead!

For example, if the SIS must reduce the rate of accidents by a factor of 100 it must not fail more than once in 100 demands. In this case it requires a PFD avg of 0.01. This term is related to the failure rate but is not the same, as we shall see later. Basically:

PFDavg = •R = 1/RRF

For our RRF example of 500 the PFDavg will be: 0.002 or 10^{-3}

Low demand vs. high demand rate. If the demand on the safety system is less than once per year the PFDavg serves as the measure of risk reduction that can be expected from the safety system.

If the demand rate is higher than once per year or is continuous, the IEC standard requires that the term to be used is the probability of a dangerous failure per hour (also known as the dangerous failure rate per hour). This measure is approximately the same as the PFDavg when taken over a period of 1 year but is applied when there is a poor chance of finding the failure before the next demand.

2.10 The concept of safety integrity level (SIL)

2.10.1 When to use an SIS and how good must it be?

The big question now is:

How do we decide when to use a safety instrumented system and just how good must it be?

The answer is:

It depends on the amount of risk reduction required after the other devices have been taken into account.

The measure of the amount of risk reduction provided by a safety system is called: The safety integrity, it is illustrated by the next diagram, from IEC.

Risk reduction layers example based on IEC 61508 Figure C1

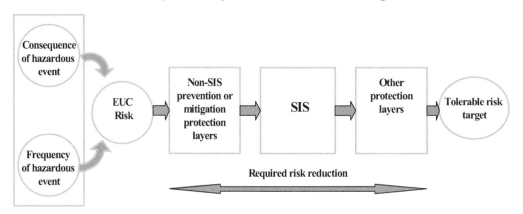

Figure 2.19
Risk reduction

Note that this diagram defines safety integrity as applicable to all risk reduction facilities. It is a general term but when it is applied to the safety instrumented system it becomes a measure of the system's performance.

In order to get a scale of performance safety practitioners have adopted the concept of safety integrity levels or (SILs). The SILs are derived from earlier concepts of grading or classification of safety systems. The principle is illustrated in the next figure where the layer of protection provided by an SIS is seen to be quantified as a risk reduction factor from which it can be converted to a PFDavg and referenced to an SIL classification table.

Determination of SIL

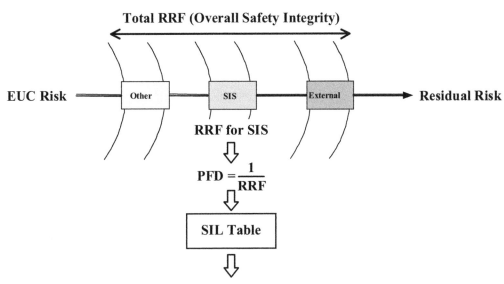

Figure 2.20
Determination of SIL

Essentially the SIL table provides a class of safety integrity to meet a range of PFDavg values. Hence the performance level of safety instrumentation needed to meet the SIL is divided into a small number of categories or grades.

The IEC standard provides the following table for SILs.

Safety Integrity Levels: based on IEC Table 2

Target failure measures (PFDavg) for a safety function operating in
a low demand mode of operation

Safety integrity level	Low demand mode of operation (average probability of failure to perform its design function on demand)
4	$\geq 10^{-5}$ to 10^{-4}
3	$\geq 10^{-4}$ to 10^{-3}
2	$\geq 10^{-3}$ to 10^{-2}
1	$\geq 10^{-2}$ to 10^{-1}

Table 2.7
Safety integrity levels

The ISA standard has a similar table and covers exactly the same range of PFD values except that SIL-4 is not used.

SAFETY INTEGRITY LEVEL	1	2	3
SIS PERFORMANCE REQUIREMENTS	Safety Availability Range		
	0.9 to 0.99		
	PFD Average Range		
	10^{-1} to 10^{-2}	10^{-2} to 10^{-3}	10^{-3} to 10^{-4}

Table 2.8
Safety integrity level performance requirements

Note that this table includes the Safety Availability Range. We can also add in the risk reduction factor, RRF as is shown in the table given in Appendix A.

An SIL 1 system is not as reliable in the role of providing risk reduction as SIL 2; an SIL 3 is even more reliable. Once we have the SIL we know what quality, complexity and cost we are going to have to consider. It should be clear from this that if we can calculate the risk reduction as a PFDavg requirement or as safety availability we can look up the tables and define the SIL we are going to need from our safety system.

2.10.2 How can we determine the required SIL for a given problem?

There are some choices about how the SIL is determined. Basically there is a choice between using real numbers (quantitative method) and some variations on fuzzy logic (qualitative methods).

One method available is to use the numerical risk reduction factor we have just been looking at, i.e. we calculate the RRF required for the safety system as recalled.

This gives us the quantitative method.

2.10.3 Quantitative method for determining SIL

Assuming that we have identified a particular process hazard during the course of a study and we wish to consider a safety instrumented system.

- The first step is to determine the risk levels. We need to set up a tolerable risk value that must be achieved or bettered (Ft)
- Then we calculate the present risk level presented by the existing situation or process (EUC risk) (Fnp)
- The ratio between the two risk levels is the factor that decides the job to be done by the overall safety system, i.e. the overall safety integrity. We can express this as the required overall risk reduction factor (Fnp/Ft)
- Now we need to determine how much of the risk reduction job must be allocated to the safety instrumented system (RRFsis). Normally we would invert this value to give the PFDavg for the SIS.
- Using the SIL table the PFDavg translates into the SIL requirement for the safety system (known as the target safety integrity).

2.10.4 Example application

Assume we have identified an EUC risk frequency of 1 per year with a consequence of > 1 injury (critical consequence). If we use a risk classification table, we shall need to reduce the risk from Class 1 to Class 3. This means that the tolerable risk Ft ranges from 1 to 500 years (0.002 events/yr) to 1 per 5000 years (0.0002 events/yr).

Taking the safest target we would try for Ft = 0.002 and this gives an overall RRF of Fnp/Ft = 1/0.0002 = 5000.

In this example we will suppose that a mechanical protection device is available as a 'non-SIS protection layer' before the SIS is needed to respond. We have estimated that it will contribute a risk reduction factor of 10. This will reduce the demand rate on the SIS by a factor of 10 and the SIS risk reduction allocation (RRFsis) becomes 500.

Using: PFDavg = 1/RRFsis, PFDavg = 1/500 = 0.002 or 2×10^{-3}

Using the IEC SIL table this gives the SIS an SIL 2 target integrity level.

2.10.5 Summary

This completes the study of the basic principles of hazards and risk reduction principles. The key points are summarized here:

- Hazards must be identified, risks are to be assessed
- The task of the safety instrumented system is risk reduction
- Separation of SIS from basic controls is generally good practice
- Layers of protection are used to provide contributions to risk reduction
- The required risk reduction defines the quality or safety integrity level of the protection system

The assessment and classification of risks is often a difficult task requiring specialist experience in the particular application being studied. A combination of historical data and careful analysis is needed to arrive at credible figures for a given risk. It is a possible stumbling block for the risk-based methodology used throughout the IEC standard.

2.11 Practical exercise

This is an appropriate stage to carry out the first practical exercise, which is intended to help clarify the tasks of risk assessment and the calculation of SIL by quantitative methods.

2.11.1 Example of SIL determination by quantitative method

Here is a simple example of how the SIL rating for a safety function can be determined by establishing the risk levels in terms of predicted frequencies before and after the application of the safety function.

In this example we suppose that a process plant has a large gas fired heater. A hazard study has identified that the combustion chamber could explode if there is a buildup of unburned gas and air after a loss of flame event. This could arise if the gas supply is not shut off as soon as the flame is lost.

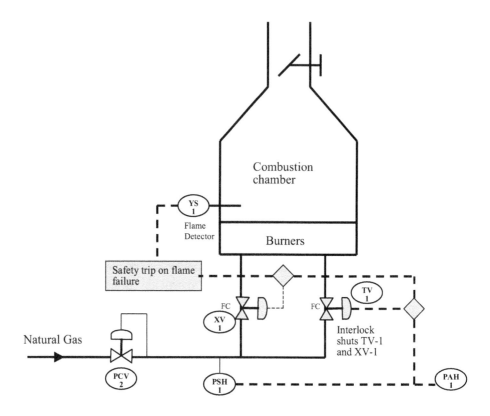

Figure 2.21
Simplified diagram of gas fired furnace

The burner controls will be fitted with a flame out detection sensor that will trip out the main and pilot gas supplies as soon as the flame is lost. This function is separate from the purge timer sequencing control system operating the ignition unit for the pilot flame. The flame failure protection system is clearly going to qualify as a functional safety system and the problem is then to determine what would be an appropriate SIL for the equipment.

Step 1: What is the risk frequency?

For the quantitative method we need to know: Fnp, the unprotected risk frequency.

How often will the flame out incident be likely to occur as an unscheduled event rather than a routine shutdown? This is the demand rate on the safety system. But how often will this lead to a furnace explosion? The project team will have to estimate this figure either from historical records or by hazard analysis methods. This is often the most difficult part of the quantitative method.

In this example we can suppose that the flame out frequency has been estimated at twice per year and that the chances of an explosion are 1 in 4 per event. This predicts the value of Fnp at $2 \times 0.25 = 0.5$ per year.

Step 2: What are consequences?

The consequence of the explosion will be a possible single fatality and extensive damage to the furnace.

Step 3 : What is the target frequency that could be tolerated?

Now we need to know what would be considered as a tolerable rate of occurrence for such an event? To do this we would consult risk profile tables for the company and also check nationally achieved accident rates of this magnitude. Let us assume it has been decided that the explosion rate should not be accepted if the chances are greater than once per 5 000 years.

This figure defines Fp, the target protected risk frequency as $1/5000 = 2 \times 10^{-4}$ per year

Step 4: Calculate the PFDavg

From the above: RRF = Fnp/Fp = $0.5 / 2 \times 10^{-4}$ = 2500

PFDavg = 1/RRF = 4×10^{-4}

Step 5: Decide the SIL

Reference to the SIL tables show that this will require an SIL 3 safety system.

The high SIL value is needed because there are no other layers of protection to partially reduce the risk. If the design team can find other measures to reduce the chances of a furnace explosion the dependency on the safety system will be reduced and probably a lower SIL rating could be specified.

Note 1:
This example shows how the SIL rating depends on the estimated risk frequency. This places great responsibility on the hazard analysis study, but this will always be a critical task in any form of risk assessment.

Note 2:
The end user determines the tolerable risk frequency in accordance with its policy on safety and asset loss evaluations. The tolerable risk frequencies have to be carefully deduced from the general targets since in each application the number of plant units per site as well as the number of persons exposed to the risk will have to be considered. See also Chapter 1 where we discussed tolerable risk criteria.

Note 3:
The demand rate found for the SIS was 2 per year even though the risk frequency for the explosion consequence was 0.5 per year. This means that the system is expected to respond to 2 demands per year. This will have to be treated as a high demand rate system under IEC 61508 although in the new process industry standard IEC 61511 this can still be considered a low demand system provided the testing frequency is high enough. The result in terms of SIL rating will be the same for either mode.

2.11.2   Comparative SILs table

This table contains failure rates for SIL values as quoted by IEC 61508 with equivalent values from ISA S84.01 and DIN 19250 (for AK values used in German standards).

IEC 61508 SIL	1	2	3	4
ISA S 84 Safety Availability	90 to 99 %	99% to 99.9%	99.9% to 99.99%	Not used
Risk Reduction Factor	10 to 100	100 to 1000	1000 to 10000	10000 to 100000
ISA S84 SIL	1	2	3	Not used
PFDavg for Low Demand rate	$\geq 10^{-2}$ to 10^{-1}	$\geq 10^{-3}$ to 10^{-2}	$\geq 10^{-4}$ to 10^{-3}	$\geq 10^{-5}$ to 10^{-4}
Failure Rate/hr for high demand rate	$\geq 10^{-6}$ to $<10^{-5}$	$\geq 10^{-7}$ to $<10^{-6}$	$\geq 10^{-8}$ to $<10^{-7}$	$\geq 10^{-9}$ to $<10^{-8}$
DIN 19250 AK values	1 to 2	3 to 4	5 to 6	7 to 8
EN 954-1 Safety categories	2 See note 1	3 See note 1	4 See note 1	Not used

Table 2. 9
General purpose table for safety integrity levels

Note 1: In machinery safety practice the European Standard EN 954 defines safety categories in terms of design characteristics for safety related controls. This is expected to be revised and issued as ISO 13849-1 and is expected to replace safety categories with a equivalents to SILs.

3

Hazard studies

3.1 Introduction

This chapter takes a closer look at those aspects of hazard studies that have a bearing on the development and specification of safety instrumented systems. It is important to understand that in general, hazard studies are a part of the overall task of safety, health, and environment management for any industrial activity, particularly in large industrial plants. Functional safety is just one part of the safety management task and hence the IEC functional safety standard supports some of the tasks of safety management but does not cover the overall task.

Examples of procedures for conducting hazard studies are shown but these serve only as an introduction to the subject and do not cover the depth of knowledge needed by an individual to conduct such studies. Specialized textbooks and training courses exist for this work, which are largely the responsibility of a process or chemical engineer. The objective here is to show how such studies are used to provide a means of identifying hazards and specifying the requirements for risk reduction.

There are four sections located towards the end of this chapter that provide additional information of value to those wishing to pursue hazard study and analysis in greater detail.

3.2 Information as input to the SRS

The starting point for an SIS is the safety requirements' specification or SRS. If we look at the input requirements for developing an SRS, we will see that much of the information needed stems from a good knowledge of the manufacturing process, its normal operations, and its potential hazards.

Input requirements for safety requirements' specification (ISA S84.01)

ANSI/ISA S84.01-1996 provides an excellent summary of the specification requirements. Section 5.2 of the standard shows the input requirements of the standard that are expected to be generated from the hazard analysis activities.

Input requirements

The information required from the process hazards' analysis (PHA), or a process design team, to develop the safety requirements' specifications includes the following:

5.2.1 A list of the safety function(s) required and the SIL of each safety function

5.2.2 Process information (incident cause, dynamics, final elements, etc) of each potential hazardous event that requires an SIS

5.2.3 Process common-cause failure considerations such as corrosion, plugging, coating, etc

5.2.4 Regulatory requirements influencing the SIS

Safety engineering personnel normally supply much of this data, except the SIL, for each function. It is preferable that an SIL classification study should be treated as an extension of hazard assessment and be handled jointly with instrument engineers.

What we find in the IEC standard is that hazard analysis and risk reduction requirements are presented as an integral part of the safety life cycle.

3.2.1 Information from hazard studies must be used

As stated by Paul Gruhn and William Cheddie in their textbook:

'It should be pointed out that identifying and assessing hazards is not what makes a system safe. It is the information obtained from the hazard analysis and the actions subsequently taken that determine system safety.'

ISA S84.01 has these useful clauses describing the hazard study stages:

'4.2.3 Once the hazards and risks have been identified, appropriate technology (including process and equipment modifications) is applied to eliminate the hazard, to mitigate their consequences or reduce the likelihood of the event. The third step involves the application of non-SIS protection layers to the process. The method(s) for accomplishing this step is outside the scope of this standard.'

'4.2.4 Next, an evaluation is made to determine if an adequate number of non-SIS protection layers have been provided. The desire is to provide appropriate number of non-SIS protection layers, such that SIS protection layer(s) are not required. Therefore, consideration should be given to changing the process and/or its equipment utilizing various non-SIS protection techniques, before considering adding SIS protection layer(s). The method for accomplishing this step is outside the scope of this standard.'

We first need to be aware of the essentials of the hazard study methods. Then we shall look at how best to interface hazard studies to the SLC activities necessary to generate a safety requirements' specification.

3.2.2 The process hazard study life cycle

A typical process safety life cycle model comprises 6 levels as shown in the following figure. The details are self explanatory.

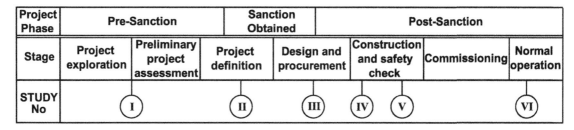

Figure 3.1
6 stage life cycle diagram for process hazard studies

Outline hazard – Study 1

- Identify hazards associated with the process
- Identify major environmental problems and assess suitability of proposed sites
- Criteria for hazards, authorities to be consulted, standards and regulations, codes of practice
- Collect/review information on previous hazardous incidents

Outline hazard – Study 2

- Examine plant items and equipment on process flow sheet and identify significant hazards
- Identify areas where redesign is appropriate
- Assess plant design against relevant hazard criteria
- Prepare environmental impact assessment

Outline hazard – Study 3

- Critical examination of plant operations also described as hazard and operability study (HAZOP) and described as FMECA, (failure mode, effect and criticality analysis)

Outline hazard – Study 4

- Reservation review verifying that the provisions in all previous studies are fully implemented

Outline hazard – Study 5

- Safety health and environmental audit (SHE) of constructed plant before introducing hazardous materials

Outline hazard – Study 6

- Final review to confirm that design has been fulfilled opposite SHE aspects
- Compare early plant operational experience with assumptions made in hazard studies
- Ensure that all documentation is available and in place

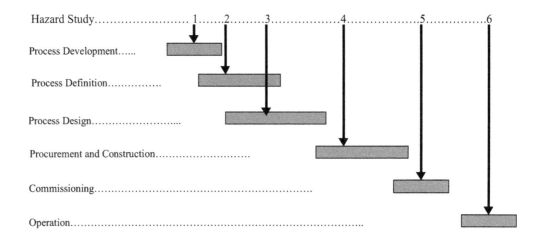

Figure 3.2
Process hazard studies at project stages

3.2.3   Alignment of process hazard studies with IEC safety life cycle

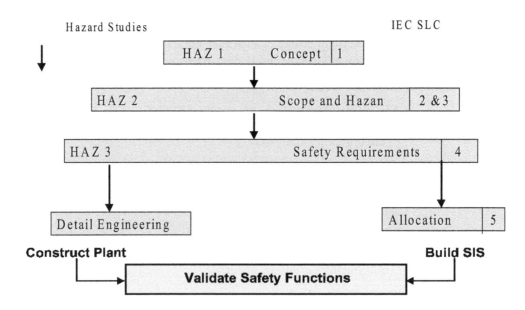

Figure 3.3
Comparing hazard studies with SLC activities

This figure shows how the IEC and ISA safety life cycle models for safety instrumented systems correspond to the established process safety life cycle models for hazard studies. The point of departure for the SIS life cycle is ideally at the end of hazard study 3 when the safety requirement's specification has been finalized.

- The plant detail engineering proceeds on the basis of P&IDs that have passed the detailed HAZOP study (hazard study 3).
- The safety instrumented system design proceeds on the basis of a safety requirements specification that has also been aligned with the detailed HAZOP study.

It seems pretty clear that the first phases of the SLC are similar in principle to the older process safety design life cycle. The main difference is the emphasis in the SLC on risk assessment in the preliminary and detailed hazard studies. This is because risk assessment leads to risk reduction, which in turn defines the SIL requirements of the safety related system.

3.2.4 History

The technique of conducting hazard studies, appropriate to the stage of design, construction or operation of a process plant, became established as a standard method in the mid 1970s through the efforts of people such as Trevor Kletz and SB Gibson working at ICI. Ref 3 in section 1.16 is a paper by SB Gibson (1975) to the UK Institute of Chemical Engineers (I Chem. E) describing the 6 levels of hazard study recommended by ICI at that time. By the late 1980s this procedure was in common use.

Trevor Kletz has become a prolific writer on the subject of hazards and safety in general and his material is entertaining and informative. See Ref. 2 in section 1.16 for a reference to one of his many publications on the subject. Most chemical manufacturing companies have incorporated hazard studies into their codes of practice for all capital projects. Hazard studies are also standard practice for modifications to existing plants and operations, unfortunately these are sometimes omitted, and this often leads to dire consequences.

Industrialized countries have requirements for risk assessment in their safety laws and in the case of hazardous processes these extend to a mandatory requirement for hazard studies to be formally carried out and recorded. Additionally the hazard studies are subject to periodic reviews to check that they are still valid for the current design of the installations, since most plants undergo substantial change throughout their lifetime.

3.2.5 Guideline documents

Some literature sources for guidance on hazard studies are given here and further references can be found in section 1.16.

IEC have recently published a very useful standard (IEC 61882) on hazard and operability studies. (See Ref. 8 in section 1.16). This standard incorporates well-established basic practices in HAZOP studies. It covers a wide range of applications such as continuous and batch processes, electronic control systems and emergency planning. It can be used very effectively as a support tool for safety life cycle activities.

The I Chem. E provides several guide documents on hazard studies, see in particular Ref 10 in section 1.16.

The American Institute of Chemical Engineers has extensive support training and literature for hazard studies, also known as process hazard analysis of PHA. See 'Publications' in section 1.16.

3.3 Outline of methodologies for hazard studies 1, 2 and 3

3.3.1 Process hazard study 1

Here is a brief expansion on typical methodologies for hazard studies: As we look at these it will be easy to see how they match up to what was described in the previous chapter for IEC 61508 phases 1, 2 and 3

3.3.2   Outline of hazard study 1

This section summarizes the method, inputs and outputs of the initial hazard study that takes place at the very early stages of a possible project.

Method

- Study team from all role players
- Systematic check of potential chemical/physical hazards against subject list

Inputs

- Process descriptions
- SHE incidents history
- Completed chemical hazard forms and data sheets
- Draft environmental and OHS statements

Outputs

- Identification of critical hazards and design constraints
- List of items requiring further action or study
- Major project decisions

3.3.3   Timing

The timing of a hazard 1 study is normally 'as soon as possible', after the initial requirement for a process or item of equipment has been identified. The study team is composed of the potential role players in the design, building and operating of the plant. The study team examines the 'concept' of the plant in the form of a simple flow diagram and outline descriptions to identify all significant hazards and environmental problems as early as possible.

3.3.4   Topics

Among the topics to be considered are:

- Project definition: e.g. business objectives, size, scope, location, timing, 0.. ....capacity
- Operating and control philosophies: e.g. 24 hr, fully automatic
- Business risks: e.g. effective of outages, breakdowns etc
- Process description: outline process, raw materials, products, effluents etc
- SHE experiences: e.g. record of other plants, known environmental problems
- Material/chemical hazards data sheets and listings
- A chemicals interaction matrix

The interactions matrix is a particularly important table because it will identify any substances that might be created by a combination of the chemicals expected to be present at the plant in question. Dangerous combinations are identified and the possibilities for their occurrence through spillages, leaks or operational errors must then be carefully considered.

3.3.5 Environmental impact

For the initial environmental impact considerations, the environmental diagram shown in Figure 3.4 is a very useful way to test and explain the overall environmental impact situation.

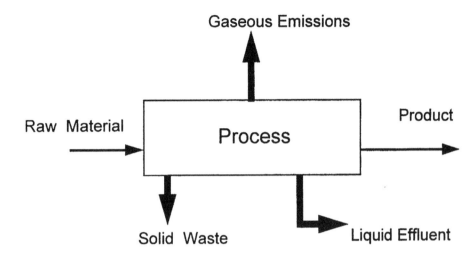

Figure 3.4
Environmental interfaces' block diagram

3.3.6 IEC: concept

Now let's see how IEC box 1 'concept' compares with the scope of hazard study 1.

Objective

- Preliminary understanding of the EUC and its environment (process concept and its environmental issues, physical, legislative etc)

Requirements

- Acquire familiarity with EUC and required control functions
- Specify external events to be taken into account
- Determine likely source of hazard
- Information on nature of the hazard
- Information on current applicable safety regulations
- Hazards due to interaction with other equipment/processes

The points shown are from clause 7.2 of IEC 61508. It is clear that using a typical hazard 1 study can satisfy 'concept.'

3.4 Process hazard study 2

Key phrases: systematic search for hazards, preliminary hazard analysis, risk reduction decisions.

3.4.1   Outline

Method

Study team from all role players

- Systematic search for specific chemical/physical hazards against plant preliminary flow sheet or block diagram.
- Select operational blocks for study. Apply keywords from a guide diagram using a sequence of questions.
- For each hazard carry out assessment of consequences and frequency.
- Use fault tree analysis where appropriate.
- Record results in a chart form.

Inputs

- Process flow sheet
- Hazard study 1 records and data sheets

Outputs

- Identification of critical hazards and design constraints
- Hazard summary table
- Risk assessment and requirements for risk reduction
- List of items requiring further action or study
- Major project decisions

3.4.2   Hazard study 2 – systematic procedure

An example of a practical method of searching for hazards, evaluating the consequences, and outlining corrective actions is given here, details are based on methods used in the AECI Process Safety Manual. (Ref 9 section 1.16).

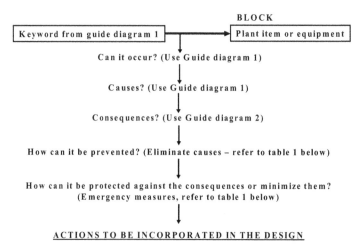

Figure 3.5
Hazard study 2 – systematic procedure applied to a process flow diagram

Use the guide diagrams and tables shown in sections 3.12 and 3.13 to follow the procedure given here:

- Take a section of any proposed plant in block diagram or preferably in flow sheet form.
- Use the keywords from HAZOP study 2 diagram 1 to run down the prompts in Guide Diagram No. 1.
- Identify any feasible hazard and record it and its possible causes in the Hazard Summary table. Some possible causes are given in the guide diagram but see also the next paragraph concerning EUC control system risks.
- For the 'consequences' use Guide Diagram No. 2 and record them in a summary sheet.
- For 'prevention' use possible ideas from experience or look for prevention layers such as those indicated in the sections or any other prevention layer technique based on experience.
- For 'protection against consequence' look for mitigation layers of protection such as those indicated in Table 3.12.

The outcomes of all the steps described should be recorded in the hazard summary table. An example of a completed entry in a table is shown in the next diagram.

HAZARD STUDY 2 RECORD	PROJECT: ETHYLENE DICHLORIDE PLANT					
W/S No:	Hazardous Event or Situation	Causes	Consequences Immediate / Ultimate	Preventative Measures (Reduce likelihood)	Protective / Emergency Measures (Reduce Consequences)	Recommendations
4	Unconfined Explosion	- ethylene supply pipeline rupture - Static sparks or nearby ignition source e.g. electrical equipment	- Damage, injuries, fatalities	- High pipeline integrity design and construction - Detection of flammable gases	- Nearby buildings e.g. control room designed not to collapse	- Confirm whether control room building design should be blast resistant

Table 3.1
Hazard Study 2: record item from ethylene dichloride plant study

Note how the columns are showing: Causes, Consequences (part of Risk), Measures to reduce likelihood (prevention), Measures to reduce consequence (mitigation). The summary sheet will form one item in a list of deliverables required from the study.

3.5 Risk analysis and risk reduction steps in the hazard study

As the hazard study, progresses there will be particular items of interest to instrument engineers or those responsible for the safety life cycle project. Here are some of them.

3.5.1   Hazards of the EUC control system

The hazard study must look for potential hazards in the EUC operating with its basic process control system (BPCS). Generally, control system failures are seen as human operator errors, loop control faults or sequencing/interlock failures. However, it is also necessary for the hazard study to consider possible failure modes of any computer/PLC systems used in the BPCS or in supervisory control. This type of study falls within the scope of what are known as 'computer hazops' or 'chazops'. Section 3.13.3 contains some introductory notes on chazops.

Examples of potential causes of failures in control systems

Listed below are some of the typical failure conditions that should be considered for the EUC control system. Included here are human operator errors.

- Human operator errors due to mistaken identities of controllers. Incorrect entry of set point values, wrong valves set to manual mode
- Networked systems may be vulnerable to external corruption
- Power failures within the network may lead to some control functions shutting down whilst others continue normally
- Software in large systems is frequently updated and it can be difficult to control
- Control failures in groups due to DCS

3.5.2   Event sequences leading to a hazard

There will be one or more logical sequences of events leading to the hazard. These must be recorded in a clear format. Fault trees are the best medium to use.

- Clarify the logic of events leading to a hazard.
- Capture the event frequencies and the probabilities of consequences.
- Model and calculate hazard event frequencies (Fnp and Fp).
- Help to develop and evaluate protection layers.

   We are going to look at fault trees and get some practice with them at the end of this chapter (see Appendix A). The point to note here is that all the possible combinations of events leading to the 'top event' should be tracked down and recorded in the hazard 2 sessions.

3.5.3   Hazardous event frequencies

Once the logic of an event has been determined the 'likelihood' or 'event frequency' has to be estimated. The hazard study team should make an initial estimate and in some cases they will call for a more detailed evaluation to be done outside of the main study. As we have seen in Chapter 2 the risk reduction requirement depends on knowing the unprotected risk frequency. Again, fault trees provide a basic platform for developing the risk frequencies.

3.5.4   Inherent safety solutions

Hazard study guidebooks advise that the design team should look first for inherent safety solutions. i.e. eliminate the hazard.

AVOID   Not having anything that is hazardous
REDUCE   Reduce the amount that is hazardous (inventory)

SUBSTITUTE   Replace with something that is not hazardous

SIMPLIFY   Use conditions that are not hazardous. (This is not always simpler!)

It looks as if inherent safety becomes part of the process design, effectively the 1st layer of protection.

3.5.5   Estimating the risk

It is the task of the hazard study team to estimate the level of risk associated with each hazard. This brings us directly back to the risk classification work we covered in Chapter 2. Recall that the quantitative method of risk analysis begins with an estimate of unprotected risk frequency and an estimate of the consequences.

Most companies develop some form of risk classification chart that serves as a guideline for the hazard 2 team to decide on acceptability of any given risk. It is most important to ensure that the risk classification information is properly agreed and approved within the company structures before the start of a hazard 2 study.

3.5.6   Adding more protection

Typically, at the point of deciding risk reduction needs, the study team will move along to the next step of suggesting more protection. Hence, the layers of protection that we saw earlier will be added. For the SLC we are going to need to recognize and classify these layers to assist in our risk reduction models.

3.5.7   Typical protection layers or risk reduction categories

The following table describes some measures as examples of risk reduction and their suggested categories are listed as follows.

A: Basic design
B: Process control
C: External risk reduction measure
D: E/E/PES safety related system
E: Other technology related risk reduction system
F: Mitigation layer

3.5.8   Key measures to reduce the risk

Once the general philosophy for prevention and mitigation has been established the hazard study team can follow up with more specific proposals. Typical key measures are listed here.

- Layout/separation distances
- Pressure relief requirements
- Safety instrumented systems (trips and alarms)
- Interlocks (e.g. mechanical, electrical, PLCs, computer, procedures)
- Fire prevention, protection, and fire fighting
- Preventing sources of ignition
- Management systems for safe operation and maintenance
- Correct operator interaction (i.e. knowing when and how to shutdown safely)
- Spillage containment and recovery absorption systems
- Reducing exposure to harmful substances

Hazard reduction measure	Category code
Inventory reduction	B
Pressure/temperature reduction	A or B
Minimize equipment, piping, seals and joints	A
Design for containing maximum pressures	A
Pressure relief systems location/layout/spacing	E
Containment/bunding/safe disposal	C
Eliminate sources of ignition	E
Rapid leak detection	F
Control operating parameters within safe limits	B
Interlocks for drives and valve settings	B or D
Monitor and alarm deviations in critical parameters	D
Shut down process on critical deviations	D
Rapid fire detection	F
Control room/occupied building blast resistant	F
Toxic refuges (gas escape rooms)	F
Fire protection	F
Dispersion aids – water jets, air dilution fire fighting facilities	F
Emergency procedures on/off site	F
Vent/relief discharges – treatment/containment/recovery	E
CHRONIC HARMFUL MATERIAL HAZARD	
Design for hygiene requirements as per OHS act	A
Containment of low-level discharges	E
Monitoring of work place atmospheres	E
On-going health screening	E
Building ventilation	A

Table 3.2
Suggested risk reduction measures

Let's consider some of the measures that will affect the SLC activities and the instrument engineers.

3.5.9   Process and operational safety measures

These are typical routine measures that are built in to the operating instructions for the plant. Operators are given standard operating procedures that include such duties as closing all drain valves and perhaps locking devices off before start up of a plant unit.

The hazop will identify many of these as low-level precautions.

3.5.10   Alarm functions

Many concerns raised in a hazop will lead to alarm functions being specified to the control systems engineer. The typical alarm functions will begin with simple deviations from the normal range, e.g. High level in a tank serves as a first warning of an approaching problem

.

- More serious alarms will merit a separate sensor for process conditions
- Smarter alarms bring rate of change warnings
- Even smarter alarms turn themselves off when not applicable

At some point in the discussion of a hazard, issues concerning response to alarms will be raised:

- Can we be sure the operator will respond?
- Will he be there?
- Is there enough time to respond?
- Will he do the right thing?
- How many other alarms will there be at the same time?

As we know all large control systems include a full range of alarm facilities and the problem is to avoid a proliferation of alarms that defeat the purpose when a real upset occurs. There have been several studies and reports on the subject of alarm overload and one of the best-known examples was the one at Milford Haven Refinery that was stated by a public enquiry to be a contributing factor to the huge explosion there in 1994.

Essentially the issue is about the management of alarms.

'On its own, one alarm is fine. A flood of alarms is a problem.'

3.5.11 Safety instrumented functions

When a hazop study finds there is a situation that is likely to arise that requires a positive and sometimes very fast response to ensure safety it will conclude that a safety instrumented function must be performed.

This decision should not be taken lightly because the cost of such functions is high and the degree of upkeep needed is not trivial. The SIS brings with it the additional factor of potential for nuisance tripping, reducing the overall availability of the plant. These considerations will not apply to applications such as burner management and compressor anti-surge protection where the need for the SIS is not in dispute. But where there is a possible choice of protection method we should consider the basic points shown here in Figure 3.6.

SIS Solutions

**When the Hazard Study team
calls for a safety- instrumented system... Remind them of the cost and
the risk of nuisance trips.**

The information needed will include:

•**Actions required**

•**Process conditions**

•**Safety times**

•**Risk assessment and risk reduction based on tolerable risk**

•**Trip logic**

Resist Prescribed Design Solutions !

Figure 3.6
SIS solutions

Sometimes a hazard study team is tempted to prescribe an apparently obvious safety function for protection against a hazard. The instrument engineer should be careful to ensure that the safety function requirements are properly defined and thought out before agreeing to proceed with the solution. This is an advantage of developing a proper safety requirements specification, as we shall see in Chapter 4.

3.6　Interfacing hazard studies to the safety life cycle

For process plant engineering, hazard study 2 traditionally plays a major role in defining the scope of hazards and influences the basic design to try to eliminate or mitigate risks.

The new safety life cycle is also required to identify specific hazards and risks in the context of the EUC, the EUC control system and all other external factors. Everything that the SLC needs for stages 2, 3 and 4 is obtainable via the hazard 2 study. Therefore, it may be helpful at this point to draw up an activity model to show how we can link up the SLC phases and the hazard studies to save everyone on a project team from a lot of duplicated effort.

(This guide diagram is included with more details in the hazard 2 guidelines in appendix B)

In this model we transfer the hazard analysis study work from hazard 2 into the SLC phase 3. The hazard study report then simply refers to the conclusions given by the SLC phase 3 work.

What if there is no hazard 2 study? Then all the work falls within the SLC. This would apply to a more limited scope of plant such as a machine tool or a burner control system.

What if the hazard analysis is fully done by the hazard study engineers? Then the results can simply be captured into the SLC phase 3 report to complete that phase of safety life cycle.

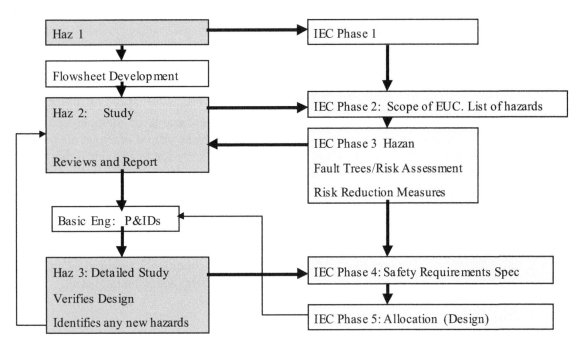

Figure 3.7
Activity model for hazard study/SLC interface

3.7 Evaluating SIS requirements

When a hazard study identifies a potential SIS requirement it will need to evaluate the feasibility and cost of the solution and decide on confirming the decision. At this point the instrument engineer must assist the hazard study team on three key issues; These are:

- Know what qualifies as an SIS? (Is an alarm part of the control system or part of an SIS? What about interlocks?)
- Know what information is required to assess the feasibility and cost of a possible SIS solution to the risk problem.
- Provide guidance on the likely design of the SIS including sensors and actuators.

Generally alarms and interlocks will be regarded as part of the BPCS unless an SIL 1 or better value is going to be claimed for them.

For assessment of the SIS feasibility the instrument engineer will need some essential information at this point. A suggested checklist is shown here.

Checklist for SIS evaluation at hazard 2 stage

- Hazard summary table.
- Fault tree or logical events leading to hazard.
- Unprotected risk frequency.
- Tolerable risk frequency (per function per unit).
- Layers of protection.
- Safe state of the process.
- Action required by the SIS to bring the process to the safe state.

- Process safety time.
- Tolerable frequency of spurious trips.

Some comments on the contents of the above list are given in the next paragraphs.

3.7.1 Tolerable risk frequency

We have already examined the meaning of this term in Chapter 2. One of the difficulties in setting this target arises when a plant has multiple hazards or multiple units. The tolerable risk frequency for the whole plant then has to be shared amongst all possible contributors to a hazardous event and the tolerable frequency for an individual risk is then much lower. The risk assessment must take this into consideration when setting a target for an individual risk

3.7.2 Safe state of the process

There is no doubt that when a trip system operates it must lead to the process arriving at a safe condition. The hazard study must define what this condition is going to be and must check that there are no other hazards associated with the trip-action taking place.

3.7.3 Trip functional requirements

These must be very carefully spelt out to minimize the chance of a systematic error at the definition stage. Fault trees and/or trip logic diagrams should be used to achieve the best possible clarity.

3.7.4 Action required to reach safe state

This information describes the output response of the SIS as seen by the process hazard study team.

3.7.5 Process safety time

IEC 61508 defines process safety time as follows:

'... the period of time between a failure occurring in the EUC or the EUC control system (with the potential to give rise to a hazardous event) and the occurrence of the hazardous event if the safety function is not performed.' Figure 3.8 shows a typical situation of rising pressure in process where there is finite time between the trigger event that indicates abnormal conditions and the hazardous event itself.

Process safety time is an important factor in specifying the response speed of the logic solver section of the SIS. (See Chapter 5.) It also has a very large bearing on the cost of trip valves and on the selection of the sensors. The hazard study team must be clear on the response time available for the SIS to act.

3.7.6 Tolerable rate of spurious trips

This defines the extent of extra protection that will be needed to avoid financial loss through spurious trips of the safety system.

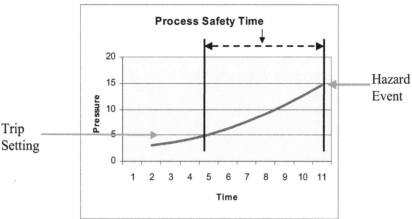

Detection, decision and actuator response must take place within process safety time

Figure 3.8
Example of process safety time

3.7.7 SIS preliminary estimate

Armed with the above information the instrument engineer should be in a position to generate a preliminary estimate of performance and cost for each specific safety function requested by the hazard study team. The contents of the report back estimate are suggested here.

For each safety function:

- Risk reduction diagram
- Safety function description
- SIL value
- Measures intended to protect against spurious trips (e.g. 2003 design)
- Requirements for sensors and actuators
- Cost estimate

3.7.8 Continuation to SRS

The information gathered for the preliminary estimate is the same as that required to generate the safety requirements specification for the next stage of the SLC. Hence, if the hazard study team accepts the initial proposals from the instrument engineer the next stage of the SLC can proceed.

3.7.9 Hazard 2 report

The initial hazard study sessions are often followed by a development phase during which the design team will improve and further specify any hazard control measures found necessary from the systematic study. This work should lead to a closing report with the following scope:

Scope of Hazard Study 2 Report:

- The hazard summary table
- Risk appraisals with quantified assessments of risks to people or the environment. Layout and spacing issues may be dealt with in this section

- Specification of key protective systems, mechanical, electrical and instrument. This will include estimates for hazard demand rates and required risk reduction factors supported by hazard analyses or fault tree logic diagrams
- Relief systems philosophy
- Service requirements in the event of an emergency
- Copy of environmental impact assessment
- Correspondence with authorities

All the above information matches the needs of the records for the IEC safety life cycle and so we can link this report directly into the SLC documentary records. We must now check the IEC requirements for phases 3 and 4 to see that we have enough information to complete the specification task.

3.8   Meeting IEC requirements

It should be clear that hazard 2 study effectively covers the IEC Phase 2: 'Overall Scope Definition.' Can it also satisfy Phase 3 of the safety life cycle?

3.8.1   IEC requirements for hazard and risk analysis

We look briefly at the IEC standard clause 7.4 that outlines the IEC Phase 3, hazard and risk analysis.

Objectives

- Define hazards and hazardous events for EUC and EUC control system
- Determine event sequences
- Determine risks

Requirements

- Hazard and risk analysis
- Probability of events, potential consequences
- Determine EUC risks
- Use qualitative or quantitative risk analysis
- Record all risk reduction data and assumptions

We can see that the IEC standard is focusing on the actual hazards and risks identified for the plant. It wants all hazards and risks to be examined and assessed with all information obtained from the earlier phases compiled into a well-documented record.

It becomes clear that the span of activities in hazard 2 covers the IEC phases 2 and 3. In the next chapter we shall see how phase 4 sets down the details of the overall safety requirements and thereby continue the interaction with hazard study activities into the hazard 3 level. It is not difficult to satisfy the need to conform to SLC requirements if we organize the interaction with hazard studies but the timing of all the activities can be a problem.

3.9   Hazard study 3

In some ways hazard 3 is simply a more detailed version of hazard 2 performed when more engineering detail has been established. But you would want to study overall hazards before looking at the details so hazard 2 is always a good precursor to hazard 3.

For an analogy: If you studied a forest for hazards you would look first at the danger of fires and the location of picnic sites before checking each tree to see if it might fall on you.

3.9.1 Outline of methodology for HAZOP

Key phrase: Line-by-Line Study of Hazards and Operability... hence the term HAZOP.

A line-by-line study of operations and deviations performed on the finished basic design by a study team. Team includes process or plant engineer, instrument engineer, commissioning engineer and end users.

Scope of application

- Continuous processes
- Batch processes
- Materials handling
- Mechanical plant and machinery
- Electrical
- Transportation/railways
- Distributed control systems. Computer controls....This item brings us to the subject of hazard studies for programmable systems

3.9.2 Outline of HAZOP method

The HAZOP method is a systematic search for specific operational hazards against plant firm (not frozen) P &I D or Eng Line Diagrams.

- Select practical sections of plant, normally a section of process pipeline linking two major vessels or operational stages. These are called 'parts for study' for example, all the plant shown in Figure 3.8.
- Within the part we can identify one or more paths where material is transferred or a change of condition is created. These are called change paths.
- The study team is prompted to consider possible deviations from design intent along each change path.
- Hazards and consequences identified as feasible are then recorded and itemized for action by the appropriate persons.

Figure 3.9 illustrates the above line-by-line method. E.g. select a path such as A-B-C-D.

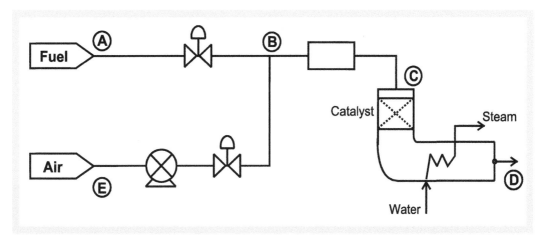

Figure 3.9
Example of a part for study with 2 possible change paths in a HAZOP study

3.9.3   Concepts of change paths and elements

Typically, a part for study will include the transfer of material from source to destination. The function of a part can then be seen as:

- Input material from a source
- Perform an activity on the material
- Product delivered to the destination.

Within this part it is possible to identify one or more change paths. It is easy to recognize a suitable change path because it becomes practical to apply 'deviations' to the operation performed along the path. Figure 3.10 shows a generalized model for the change path concepts.

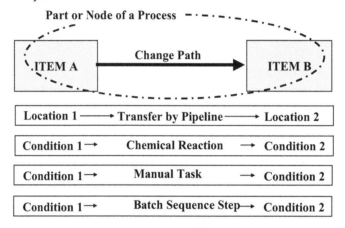

Figure 3.10
Change path concepts

Physical movements from A to B or condition changes from one substance to another both qualify as changes. A single step of a batch manufacturing operation produces a new location or new condition. (In batch control work these are often called 'state changes'.) The operation performed to create the change is the change path and this where the deviations from design intent will be applied in the examination. Fig 3.11 shows deviations applied to the change path and asks for possible causes and consequences

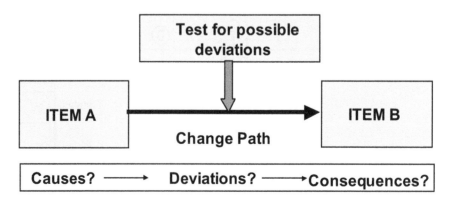

Figure 3.11
Deviations applied to the change path

If this seems rather detached from reality try creating change path applications for real life examples. In Figure 3.12 we have tried an operational step in assembling a rotary lawn mower. The change path is to fasten the blade to the shaft on the mower.

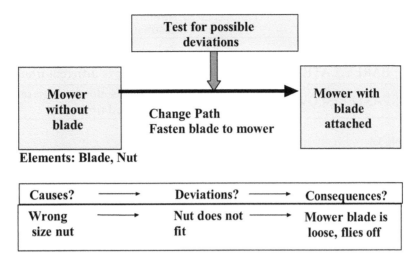

Figure 3.12
Change path example for assembly task

In the above example a change of state is intended in the mower. The task of fastening the blade is the change path and having the wrong size of nut causes the deviation. The elements involved in this operation are the worker, the blade and the nut. It is the elements of the task that are capable of deviating.

3.9.4 Generating deviations

We know that a deviation is to be considered for an element or parameter. The most common types of deviation can be listed as a set of guidewords. A common guideword list is used for HAZOP work in any particular industry sector with additional guidewords being available to help stimulate other possibilities in the minds of the team.

The guideword system begins with a set of basic guidewords that will always apply to any element. The basic guidewords and their generic meanings are shown in the following table.

Guideword	Meaning
NO or NOT (or none)	None of the design intent is achieved
MORE (more of, higher)	Quantitative increase
LESS	Quantitative decrease
AS WELL AS (more than)	Qualitative modification or additional activity occurs
PART OF	Only some of the design intent is achieved
REVERSE	Logical opposite of design intent
OTHER THAN	Complete substitution – another activity takes place

Table 3.3
Basic guidewords and their meanings

Some other commonly used guidewords are shown in Table 3.4

Guide word	Meaning
WHERE ELSE	Applies to flows, transfers, sources and destinations
BEFORE/AFTER	Relates to order of sequence
EARLY/LATE	The timing is different from intention
FASTER/SLOWER	The step is done faster or slower than the intended timing

Table 3.4
Guidewords relating to location, order or timing

The basic guidewords lack any real meaning until they are combined with elements or characteristics within elements.

Table 3.5
Guidewords relating to location, order or timing

Combining guidewords with elements generates a matrix of deviations, some of which are credible and some are not credible. It therefore falls to the study team to decide which deviations from the matrix they are going to consider.

Derived guidewords

Here is a table based on the HAZOP guide published by UK I Chem. E for typical derived guidewords generated by the parameter and guideword combinations for some process parameters. This table makes it easier to visualize the possible deviations.

Parameter	Guidewords that can give a meaningful combination
Flow	Non-, more of, less of, reverse, elsewhere, as well as
Temperature	Higher, lower
Pressure	Higher, lower, reverse
Level	None, higher, lower
Mixing	Less, more, none.
Reaction	Higher (rate of), lower (rate of), none, reverse, as well as
Phase	Other, reverse, as well as
Composition	Part of, as well as
Communication	None, part of, more of, less of, other, as well as

Table 3.6
Guidewords applied to process parameters yield real deviations

Generating the derived guidewords such as those shown above is part of the team leader's responsibility in each study session. In summary we see the relationship between guidewords and the parts to be studied as shown in Figure 3.14 below which is based on diagrams used in the AECI HAZOP manual.

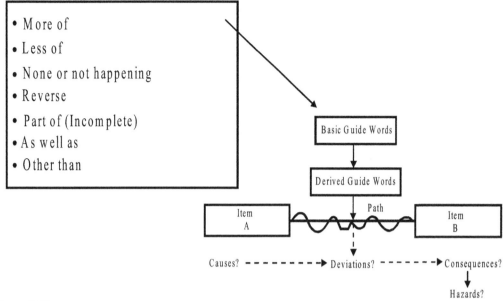

Figure 3.13
Application of guidewords to the change paths finds deviations that may have hazardous consequences

3.9.5   Study procedure

The hazard/study team leader or facilitator has the task of taking the team members methodically through a sequence of questions for each recognized deviation. Figure 3.14 shows the sequence of questions that should be asked of the team.

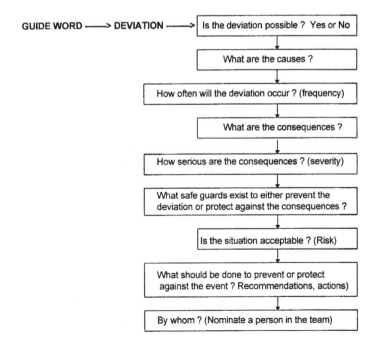

Figure 3.14
Logical steps in the processing of each deviation

The sequence shown above will naturally change in nature according to the problem but it represents the core of the procedure that has to be followed for each legitimate deviation.

3.9.6   Causes of deviations

If the plant design is basically sound the cause of a deviation will nearly always be due to a failure of some kind: The following categories are common;

Hardware: Equipment, piping, instrumentation, design, construction, materials
Software: Procedures, instructions, specifications
Human: Management, operators, maintenance
External: Services (steam, power), natural (rain, freezing), sabotage.

3.9.7   Consequences of deviations

The consequences of a deviation may be an operational problem or it may be a hazard or minor or major concern. In our field of interest we are going to be interested in those hazards which justify a safety instrumented system or which call for a trip device or ESD system. The duty of the HAZOP team is to accurately state what is known about the consequences including the possible hazards that may arise – e.g. tank overflow may be trivial if it is water but disastrous if it is a strong acid or if it is flammable liquid that could create a fire.

3.9.8   Adding protection layers

The application of protection layers by the HAZOP study team may have already been done in the hazard 2 stage. If any remaining problems are found at the detail level then more detailed protection layers may be proposed at this stage.

Some typical solutions in the chemical industry can be found from the following list.

- Provide a relief system in the piping (overflow, return path, vacuum breakers)
- Provide manual isolation valves or non-return valves
- Standby pumps, auto starting on alarm or stop of the first pump
- Mechanical restrictors against high flows
- Reduce exposure of operators hazard (reduces C rather than freq)
- Additional actions sequenced in the process control system
- Alarms on process controller excursions to sound in control room
- Back up/independent alarms/louder alarms
- Simple interlocks (e.g. pump to trip on low level before it runs dry)
- Safety trip devices
- Chemical bombs to kill reactions
- Emergency cooling
- Gas detectors
- Fire deluge systems
- Blast walls
- Ejector seats!

Many of the basic non-SIS solutions are built into processes through the experience of the equipment or process engineers before the HAZOP study is started. Any proposed safety instrumented functions will be reviewed at this stage. The HAZOP study validates these solutions and records them and often adds several more good basic safety solutions.

The problem for the safety life cycle team is that each new safety measure may require an update to the documentation already drafted for the safety requirements specifications.

3.9.9 Recording of HAZOP results and safety functions

As a hazard study progresses the identified items are often recorded in the form of an itemized table with fields set up as follows:

Project:			Path	Intention	
Deviation	Causes		Consequences	Action	Ref no
No flow	Blocked suction at pump		Overheating fire in pump	Low flow to trip pump after 10 seconds	1

Figure 3.15
Pro forma of record for HAZOP results

One of the problems associated with the final hazop studies is that new SIS functions are often introduced at a late stage of the plant design. These additions must be captured into the existing records for the SLC phases 3 and 4. Hence the draft hazard and risk analysis will have to be updated along with the SRS and safety allocation records. It is the task of the person responsible for managing the SLC to see that all new additions are properly recorded.

Suggestion: Maintain a strictly safety functions register supported by a reference numbering system.

3.10 Conclusions

We conclude that hazard studies do most of the work needed for the safety life cycle. This work must be done to provide a valid way forward into the next phase, which is the safety requirements specification.

If it is your responsibility to set up a functional safety record for a plant you will need to see that the information required by the IEC standard is provided for all risk reduction items. In some cases this phase may have to be re-visited and updated as more information becomes available.

Suggestion: If a hazard study life cycle is used the instrumentation team involved in using ISA or IEC standards may need to assist or encourage the plant engineers to generate the hazard and risk analysis as an extension to the basic HAZOP study.

Suggestion: Instrument and control engineers can use a checklist during the hazard studies to ensure that all the data they are going to need for the safety life cycle records has been obtained and agreed to with their colleagues.

3.11 Fault trees as an aid to risk assessment and the development of protection schemes

3.11.1 Fault trees

- Help to clarify the logic of events leading to a hazard
- Capture the event frequencies and the probabilities of consequences

- Are used to model and calculate hazard event frequencies (Fnp and Fp)
- Can be very useful for evaluating protection layers

Introduction to fault tree analysis

Fault trees are a valuable aid to risk assessment and the development of protection schemes. They are normally used once the HAZOP study has identified a potentially hazardous event and the team has requested some analysis of the likelihood and consequences.

According to an Instrument Society of America (ISA) technical report (ISA Tr84.03) FTA was originally developed in the 1960s to evaluate the safety of the Polaris missile project and was used to determine the probability of an inadvertent launching of a Minuteman missile. The methodology was extended to the nuclear industry in the 1970s for evaluating the potential for runaway nuclear reactors. Since the early 1980s, FTA has been used to evaluate the potential for incidents in the process industry, including the potential for failure of the safety instrumented system (SIS).

FTA begins with a 'top event' that is usually the hazardous event we are concerned with, for example, 'explosion'. The 'tree' is then constructed by developing branches from the top down using two basic operators: 'AND gate' and 'OR gate'. The logic gate symbols are shown in Figure 3.16.

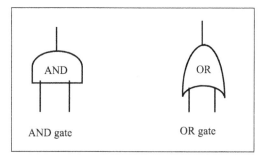

Figure 3.16
Logic gate symbols for FTA

The logic gates allow the contributing causes of the top event to be set out and combined according to the simple rules of AND gates and OR gates as shown in Figure 3.17.

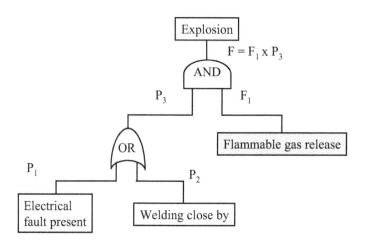

Figure 3.17
Example of simple AND and OR gate functions

Functions of the gates

AND gates are used to define a set of conditions or causes in which all the events in the set must be present for the gate event to occur. The set of events under an AND gate must meet the test of 'necessary' and 'sufficient'.

'Necessary' means each cause listed in a set is required for the event above it to occur; if a 'necessary' cause is omitted from a set, the event above will not occur.

'Sufficient' means the event above will occur if the set of causes is present; no other causes or conditions are needed.

OR gates define a set of events in which any one of the events in the set, by itself, can cause the gate event. The set of events under an 'OR' gate must meet the test of 'sufficient'.

The information about each event is described as either:

P = Probability of the event occurring, or
f = Frequency of the event, or
$f \times t$ = Duration of the event.

From the above parameters the following combinational rules are obtained:

Inputs	Gate	Operation	Output of the gate
P_1, P_2		$P_1 \times P_2$	P
P_1, f_1	AND	$P_1 \times f_1$	f
$(f_1 \times t_1), (f_2 \times t_2)$		$(f_1 \times f_2)(t_1 + t_2)$	f
P_1, P_2		$P_1 + P_2$	P
f_1, f_2	OR	$f_1 + f_2$	f

Table 3.7
Combinational rules for fault trees

The combinational rules allow the information known about each individual event to be combined to predict the frequency of the top event and the intermediate events.

Event symbols

Event symbols used in FTA are shown below.

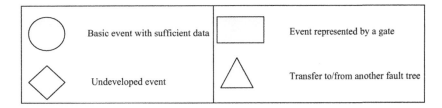

Figure 3.18
Event symbols

These provide a means of classifying events.

- A basic event is the limit to which the failure logic can be resolved. A basic event must have sufficient definition for determination of an appropriate failure rate.

- Undeveloped events are events that could be broken down into sub-components, but, for the purposes of the model under development, are not broken down further.
- An example of an undeveloped event may be the failure of the instrument air supply. An undeveloped event symbol and a single failure rate can be used to model the instrument air supply rather than model all of the components. FTA treats undeveloped events in the same way as basic events.
- Rectangles are used above gates to declare the event represented by the gate.
- Transfer gates are used to relate multiple fault trees. The right or left transfer gates associate the results of the fault tree with a 'transfer in' gate on another fault tree.
- House events are events that are guaranteed to occur or guaranteed not to occur. House events are typically used when modeling sequential events or when operator action or inaction results in a failed state.

The fault tree construction proceeds by determining the failures that lead to the primary event failures. The construction of the fault tree continues until all the basic events that influence the Top Event are evaluated. Ideally, all logic branches in the fault tree are developed to the point that they terminate in basic events.

In modeling for safety instrumented systems the basic events will typically be the failures of a sensor and/or the occurrence of an initiating event in the process operations. Figure 3.19 provides an example of an FTA model for a batch process with potential instrument and process equipment failures.

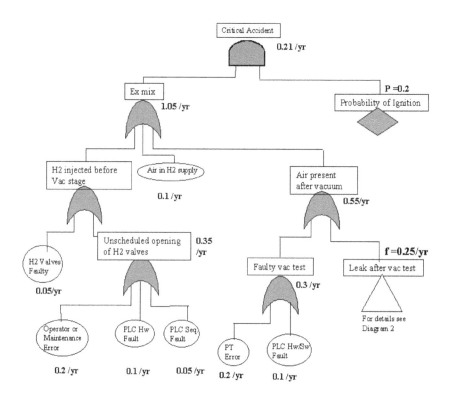

Figure 3.19
Fault tree for explosion in a batch process operation shown without SIS protection

This example is based on a batch processing plant where a PLC is used to sequence the operations of a large pressure vessel. During the batch cycle, a stage is reached where hydrogen must be supplied into the vessel and its pressure must be gradually raised to 6 bar.

The problem is that initially a large vapor space filled with air is in the vessel. To avoid a dangerous gas mixture the air must first be evacuated. There is also the possibility of leaks into the vessel whilst it is under vacuum. The PLC therefore carries out a leak test stage before starting the hydrogen feed. This test also reduces the risk of hydrogen leaks from the vessel once the pressure is positive.

The EUC risk (IEC terminology) is then due to a range of mechanical defects and PLC control/instrument defects. The fault tree shows how these defects are grouped under certain types of hazards all leading up to the potential for an explosion. The failure rates are estimated values but they do comply with the IEC requirement that normal control systems should not be credited with a failure rate lower than 10^{-5} per hr.

Summary of rules for constructing fault trees

From an internal ICI Publication: 'Guide to Hazard Analysis' by J L Hawksley.
Aim: To do as simple a study as possible i.e. get the maximum benefit from the minimum of work.

This requires: the correct logic, the most significant causes to be identified, a 'broad brush' fault tree to be drawn initially and only the areas of significant concern developed in greater detail.

Requirements.

- A physical description of the system
- A logical description of the system
- A clear definition of the hazard of concern
- A plant visit (if the system exists)

Key points for drawing up the fault tree:

- The fault tree is composed of events joined by 'AND' and 'OR' gates
- Start with the hazard or 'top event'
- Draw a demand tree first, and then add the protective systems
- Think in small steps
- Think in terms of physical properties and relative physical positions of equipment
- Check very carefully for common mode effects
- There is not necessarily a single 'correct' fault tree
- Have a box containing a reasonable and adequate description at every gate
- Check the logic of a completed tree by going along each branch to the top event

Adding risk reduction measures in FTA

So far we have seen how useful fault tree analysis can be for analyzing the risk of a known top event. The next step is to build in the possible risk reduction measures and predict the new risk frequency for the top event. It's easy to do this using the general approach shown in the next diagram.

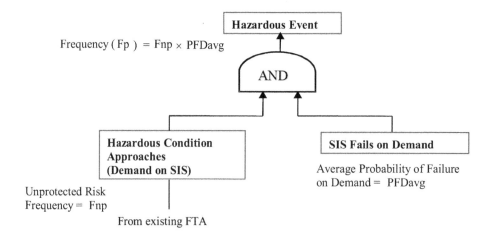

Figure 3.20
Generic method for adding SIS protection into a fault tree model

In the diagram the protection in the form of an SIS has a proposed probability of failure on demand (PFD_{avg}). This probability feeds into an AND gate with the predicted frequency of the hazardous event (demand rate F_{np} per yr) assuming no protection. The additional AND gate reduces the frequency of the hazard to $F_{np} \times PFD_{avg}$.

Event tree analysis

Event tree analysis is another widely used modeling method that is closely related to FTA. In effect this method takes the top event and divides the possible consequences into as many options as may exist and defines the probable splits.

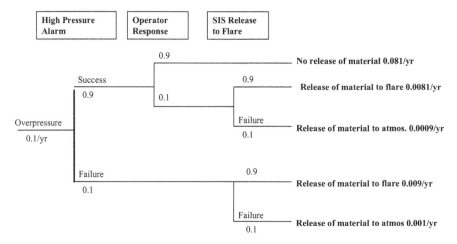

Figure 3.21
Event tree analysis example

In the event tree diagram the success or failure of the protection responses defines the choice of route to the end event. This clearly provides a consequence tree and uses the PFD of each protection measure to calculate the frequencies of each outcome.

3.12   Hazard study 2 guidelines

3.12.1   Introduction

This guide outlines a typical method for carrying out a process hazard study level 2 for a new or existing chemical process plant. It is based on guidelines originally prepared by IC I in the UK. For a more comprehensive treatment of hazard study procedures reference should be made to the hazard study guides listed in this book.

The method outlined here has been modified to cover the activities required for running the study in association with a functional safety life cycle styled on IEC 61508.

Hazard study 2 has the objectives of identifying significant hazards, their possible causes and protective measures to meet relevant criteria laid down in hazard study 1. The study is normally carried out on a new process design as soon as detailed flow sheets have been proposed and at stage before design changes are likely to incur significant costs for rework.

When applied to existing plants, hazard study 2 has similar objectives and the study is usually carried out to identify aspects of existing hardware and operations that do not comply with modem standards or criteria. Critical features of the hardware, procedures and instructions will be identified.

3.12.2   Method

Preparation

The study team initially reviews actions brought forward from hazard study 1, and clarifies key principles of operation, which will assist in carrying out hazard study 2, and which should be included in the project specification.

Systematic study

The study team systematically works through sections of the flow sheet, identifying significant hazards associated with each stage of the process (often associated with loss of containment 'LOC'). Consideration should also be given to non-process activities, which can give rise to potential safety, health or environmental hazards.

For each new diagram or system, a brief explanation of the process and proposed operation is required. Where there are several unit operations, the team can study them as a small number of separate 'blocks'. The hazard study leader will provide guidance in choosing a suitable division into these 'blocks'.

The study then proceeds, considering one block at a time, using the keywords from the attached hazard study 2 guide diagrams 1 and 2. (Figure 3.23. and Figure 3.24.)

Outcomes of the study should be recorded on the 'Hazard Study 2 Summary Table'. The summary table will be developed during hazard study 2 and subsequent studies to provide a concise overview of the hazard control measures. This information should form the basis for phases 3 and 4 of the safety life cycle records for the project.

Basis for safety

Having carried out hazard study 2, the 'Basis for Safety', including health and environmental aspects, may be established and recorded in the project specification.

Key measures include:

- Layout/separation distances
- Pressure relief requirements
- Functional; safety systems (instrumented protective systems) and alarms

- Interlocks (e.g., mechanical, electrical, PLCs, computer, procedures)
- Fire prevention, protection, and fire fighting
- Preventing sources of ignition
- Management systems for safe operation and maintenance
- Correct operator interaction (i.e., knowing when and how to shutdown safely)
- Spillage containment and recovery absorption systems
- Reducing exposure to harmful substances

These factors should then provide the basis for:
- Pressure relief reviews
- Functional safety requirements specification (to be covered in the SLC procedures)
- Identification of critical machine systems
- Fire reviews
- Electrical area classification
- Operating and maintenance policies and procedures
- Operating and maintenance information and instructions
- Conforming with major hazard regulations
- Conforming with health exposure regulations
- Computer control specification and computer security requirements (consider the checklist in section 3.14)
- Biological hazard classification (leads to definition of containment principles for materials with biological hazard)

3.12.3   Review of hazard study 2

Review meetings

Review meetings will often be arranged and chaired by the project manager or his nominated deputy. It is usually best before arranging a first review meeting to have completed the major part of the actions required, and certainly any major changes of the design. There is, however, a stimulant effect on project teams of knowing that a review meeting is imminent and this may help to complete the hazard study more promptly.

The hazard study leader should help the team to check that the issues raised by the hazard study team have been properly satisfied.

Functional safety issues

The hazard 1 study may have decided that functional safety items such as hazards arising from operation of the process or its equipment should be managed in accordance with the principles laid down in IEC 61508. In such cases the hazard study team will be directed to provide inputs to the Safety Life-Cycle (SLC) Phase 2 and Phase 3 studies and to provide continuing support for the SLC studies.

Review meetings will then include the results of SLC study work, which will provide feedback on the risk assessment and proposed risk reduction measures associated with functional safety problems.

The team, guided by the hazard study leader, needs to consider:
- Does the solution deal with the concern?
- Does the solution introduce new concerns?

Some new hazard study work is often necessary if significant changes to the design or operation are involved in the proposed solutions.

It is normally a function of the project manager or his nominated deputy to notify the hazard study leader of any changes in the design or operation that have been made subsequent to the hazard study. They can then decide whether it is necessary to hold a further hazard study meeting to consider the changes.

3.12.4 Hazard study 2 report contents

General Information

This section should contain or refer to the project specification, the flow sheet and other documentation that was studied, and essential correspondence and information relating to hazard study 2.

Hazard summary pro-forma

Hazard summary sheets (as shown in Table 3.11) provide a framework for the safety report preparation (where appropriate), future training, and auditing.

Performance against criteria

Summary of key functional safety requirements, estimated demand rates and risk reduction factors together with target safety integrity levels (SILs), analyses and fault tree logic diagrams where appropriate.

Actions identified in hazard study 1

Confirmation that necessary information has been obtained or is no longer needed.

Risk appraisals

Layout and spacing considerations as developed during hazard study 2. Also any special building design requirements, area classification requirements, and quantified risk assessments, overpressure circles or offsite risk analyses as appropriate.

Environmental impact statement

Correspondence with authorities

Statutory approvals

Copies of relevant correspondence, conclusions and agreed recommendations to be adopted.

Special health hazards

Following on from the identification stage at hazard study 1, the report should contain details of how materials requiring special handling precautions and operating principles will be dealt with. Results of exposure assessments appropriate for the COSEH regulations should be included.

Relief philosophy

Special major hazard aspects which dictate philosophy of the relief systems review should be included in this report.

Service requirements

Overview of requirements including safe shutdown in an emergency and list of services, dependencies, and design specifications.

Documentation of review meetings

3.12.5 Diagrams and tables supporting hazard study 2

The following diagrams and tables are included as typical guideline material for hazard studies

- Figure 3.22, Systematic study guide diagram 1
- Figure 3.23, Hazard study 2, guide diagram 1: identification of hazards
- Figure 3.24, Hazard study 2, guide diagram 2: identification of consequences
- Figure 3.25: Activity model: Hazard studies and SLC
- Table 3.9: Measures to prevent or eliminate causes
- Table 3.10: Measures to mitigate or reduce consequences
- Table 3.11: Pro-forma hazard summary table.

Hazard Study 2 Guide Diagram 1

Hazardous Event / Situation	Prompts	
EXTERNAL FIRE	FUEL	Flammable gas, vapor, solid, metal, wood, waste material, pyrophoric material
	RELEASE MECHANISM	L.O.C., poor housekeeping
	IGNITION	Sparks, flares. static, friction, vehicles, hot spots, welding, lightning, auto-ignition, furnaces
INTERNAL FIRE (in equipment)	FLAMMABLE MIXTURE	Flammable gas, vapor, solid, metal, dust, residue, pyrophoric material, oxygen, halogen
	IGNITION	Sparks, static, friction, welding, decomposition
INTERNAL EXPLOSION (in equipment)	PHYSICAL OVERPRESSURE	L.O.C. (Burst-Physical overpressure), head pressure, liquid filling, testing, purging
	UNCONTROLLED REACTION	Runaway reaction, decomposition, polymerization, contamination
	FLAMMABLE MIXTURE	Flammable gas, vapor, liquid, solid, dust, mist, oxygen, halogen, NC13, explosive/unstable compound, polymerization, loss of ignition/re-ignition
	IGNITION	Sparks, static, friction, hot spots, welding, decomposition
CONFINED EXPLOSION (in building)	FLAMMABLE MIXTURE	Flammable gas, vapor, dust, mist, oxygen enrichment, halogen, explosive/unstable compound
	RELEASE MECHANISM	L.O.C., storage, handling
	IGNITION	Sparks, static, friction, welding, machines, vehicles, hot spots
UNCONFINED EXSPLOSION	FUEL	Flammable gases, vapors, dusts, mists, explosives/unstable compounds
	RELEASE MECHANISM	L.O.C., storage
	IGNITION	Sparks, flares, static, friction, vehicles, hot spots, welding, lightning, furnaces, pylons
ACUTE HARMFUL/NOXIOUS EXPOSURE	ACUTE HARMFUL/NOXIOUS	Toxic gases, vapors, mists, liquids, dusts, fumes, acids, alkalis, biological
	EXPOSURE MECHANISM	L.O.C., decontamination, mechanical handling, sampling, manipulation, ventilation failure
CHRONIC HARMFUL/NOXIOUS EXPOSURE	CHRONIC HARMFUL/NOXIOUS	Toxic gases, vapors, mists, liquids, dusts, fumes, biological, radioactive
	EXPOSURE MECHANISM	Leaks from seals, valves, charging, discharging, preparing for maintenance, loading, packing
POLLUTION	POLLUTANT	Aqueous, gaseous, ground, silts, smells, fire, water, surfactants, lubricants, foams, acids, algae, flue gases, by-products, residues, dust, mists, steam
	RELEASE MECHANISM	L.O.C., decontamination
VIOLENT RELEASE OF ENERGY	ENERGY SOURCE	Electrical, potential, kinetic
	RELEASE MECHANISM	Electrical explosion, L.O.C. (burst), impact, mechanical failure
NOISE	SOURCES	Machinery, ejectors, flares, pressure let-down, vents, reliefs, road/rail traffic, sirens, conveyors, mechanical handling, demolition, construction
VISUAL IMPACT	APPEARANCE	Building profile, stacks, layout, colour, location, smoke, steam, plumes, flares, flashes
MAJOR FINANCIAL EFFECT	FACTORS TO CONSIDER	Loss of business, business interruption, downtime, overhauls, loss of services, spares, major downtime, e.g. computers, single stream equipment, etc., major operating costs

Figure 3.22
Systematic study guide diagram 1

Hazard Study 2 Guide Diagram 2

Hazardous Event/Situation	Immediate consequences	Ultimate consequences	Ultimate consequences codes
EXTERNAL FIRE e.g. Pool fire, flash fire, torch fire, BLEVE, fireball, lagging fire, electrical fire	**ENGULFMENT** **THERMAL RADIATION** **FIRE DAMAGE/FIRE FIGHTING EFFECTS** **SMOKE KNOCK-ON EFFECT***	A, E, F, G, J, M A, E, F, G F, G, I, K A, C, E, H, J, K, L	**EMPLOYEES** A: Injuries / fatalities B: Ill-health / long-term fatality **PUBLIC** C: Injuries / fatalities D: Ill-health / long-term fatality
INTERNAL FIRE (in equipment)	**FIRE DAMAGE POLLUTION (FUMES, GASES) KNOCK-ON EFFECT***	F, G A, C, E, H, J, L	**FIRE-FIGHTERS** E: Injuries / fatalities to fire fighters **PLANT DAMAGE**
INTERNAL EXPLOSION (in equipment) e.g. Contained explosion, relieved explosion, burnt containment, detonation & **CONFINED** **EXPLOSION** (in building/ structure) or **DETONATION**	**EQUIPMENT DAMAGE** **MISSILES/ FRAGEMENTATION** **L.O.C.** **BLAST DAMAGE** **KNOCK-ON (PRESSURE FILING?)*** **STRUCTURAL DAMAGE**	A, C, E, F, G, J, M A, C, E, F, G, J, M A, C, E, F, G, J, M	F: Damage to plant & equipment G: Loss of production **ENVIRONMENTAL DAMAGE** H: Harm to flora & Fauna I: Fish kill **PUBLICITY / MEDIA** J: Bad publicity K: Public / product concern **AUTHORITIES** L: Environmental protection M: Industrial incidents/accidents
UNCONFINED EXPLOSION e.g VCE, explosion, detonation	**OVERPRESSURE EFFECTS** **LOUD NOISE** **MISSILES** **KNOCK-ON-L.O.C.***	A, C, E, F, G, J, M A, C, E, F, G, J, M	**OTHER EFFECTS** N: Evacuation of site O: Evacuation of public P: Obnoxious odour * Consider possible release or loss of radio-active material
ACUTE HARMFUL/NOXIOUS EXPOSURE	**ACUTE EFFECT ON EMPLOYEES** **ACUTE EFFECT ON PUBLIC**	A, M C, E, M	
CHRONIC HARMFUL/NOXIOUS EXPOSURE	**CHRONIC EFFECT ON EMPLOYEES** **CHRONIC EFFECT ON PUBLIC**	B, M D, M	
POLLUTION	**VISUAL (inconvenience / disturbance)** **HARM TO FLORA AND FAUNA** **FISH KILL**	J, L H, I, J, K, L I, J	
VIOLENT RELEASE OF ENERGY	**EQUIPMENT DAMAGE** **KNOCK-ON-L.O.C.** **MISSILES**	F, G, M A, C, E, F, G, J, M	
NOISE	**NUISANCE / HAZARD TO EMPLOYEES** **NUISANCE / HAZARD TO PUBLIC**	J, K	
VISUAL IMPACT	**NUISANCE / ANNOYANCE TO PUBLIC**	J, K	
MAJOR FINANCIAL EFFECT	**DIRECT DAMAGE** **CONSEQUENTIAL LOSS**	F, G G	

Figure 3.23
Guide diagram 1

HAZARD STUDY 2, GUIDE DIAGRAM 2, SHEET 1

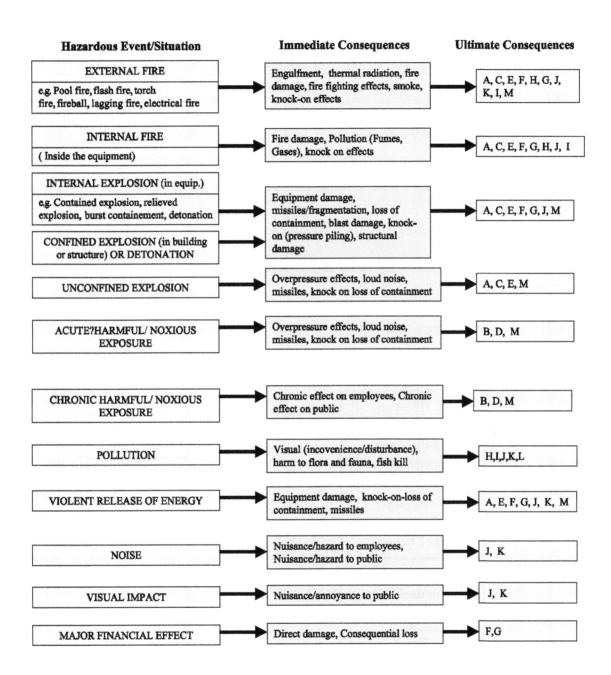

Figure 3.24
Guide diagram 2

Code	Group	Consequences
A	Employees	Injuries/ fatalities
B		Ill health/ long term fatalities
C	Public	Injuries/ fatalities
D		Ill health/ long term fatalities
E	Fire fighters	Injuries/ fatalities
F	Plant Damage	Damage to plant & equipment
G		Loss of production
H	Environmental Damage	Harm to Flora & Fauna
I		Fish Kill
J	Publicity / Media	Bad Publicity
K		Public/ product concern/site license
L	Authorities	Environmental protection
M		Industrial incidents/ accident investigators
N	Other Effects	Evacuation of site
O		Evacuation of public
P		Obinoxious odour

Table 3.8
Consequence guide

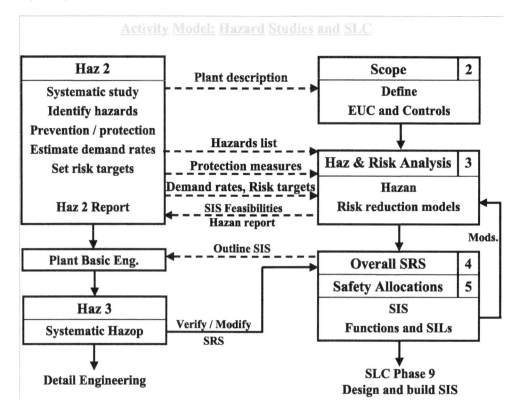

Figure 3.25
Hazard studies and SLC

Measure to prevent causes or mitigate consequences

The identification of measures to reduce risk takes place during the hazard study 2. It is useful for the study team to have a set of prompts of typical measures available. The best measures are those that prevent the causes of hazards as given in Table 3.9.

Measure	Reduce Hazard due to
Pressure/temperature reduction in process	**High energy levels, stresses**
Minimize equipment, piping, seals and joints	Leaks
Design for containing maximum pressure	**Rupture/bursting**
Provide pressure relief system	**Rupture/bursting**
Location/layout/spacing	**Interactions/confined spaces**
Operational alarms	**Wrong operating conditions**
Automatic protection systems (SIS)	**Wrong operating conditions, dependency on human response**

Table 3.9
Measures to prevent or eliminate causes

Measures to reduce consequences are used when the causes of a hazard cannot be further reduced. These measures accept that the hazardous event may occur but provide means of mitigating the scale of events or reduce the consequences. Examples are given in Table 3.10.

Measure	Mitigate Consequences of
Containment/bunding/safe disposal	**Uncontrolled dispersion, contamination**
Rapid leak detection	**Leaks leading to gas clouds/liquid pool**
Rapid fire detection	**Runaway fire**
Control room/occupied buildings design for pressure shocks	**Injury to occupants**
Toxic refuge (Gas safe room)	**Toxic vapour exposure**
Fire protection/dispersion aids – water jets	**Spread of fire**
Fire fighting facilities	**Uncontrolled fire**
Off site vent/Relief discharges	**Uncontrolled emissions**
Isolation of stages and units	**Migration of fires** **Feeding of fires from other units**
Emergency procedures	**Uncontrolled responses** **Chaotic evacuation**
Emergency shutdown systems	**Slow response to hazardous event** **Dependency on human factors**

Table 3.10
Measures to mitigate or reduce consequences

Chronic harmful material hazard: Design for hygiene standards, containment of low-level discharges, monitoring of work-place, on-going health screening, building ventilation.

Hazard Study 2 Summary Form	Project:			Drawing Number	Rev Nos.
Team Members:				Date:	Sheet No.
Hazardous Event or Situation	Caused by/ sequences of events	Consequences, immediate/ ultimate	Estimated likelihood, Suggested measures to reduce likelihood	Emergency measures (to reduce consequences)	Action required

Notes

Hazardous Event:	Each possible event is recorded as it is identified by the study team (use of guide diagram 1 or from any other prompts)
Caused by/ sequence:	Possible causes typically as shown in guide diagram 1. Details of the sequence of events that could create the conditions.
Consequences:	Posssible consequences as indicated by guide diagram 2 or from specific analysis of the event. Scale of consequences will help decide risk reduction requirements
Estimated likelihood:	Essential to have an estimate of frequency of event or an agreed descriptive term that translates to a frequency band.
Measures to reduce likelihood:	These are the suggested layers of protection and may include an SIS function.
Emergency measures	Actions to be taken once the event occurs. Also known as migration layers, these should reduce the ultimate consequences of the event.
Action required	Requirements for further study or for the next steps in the design of the safety measures.

Hazard 2 study Reporting Form:	Project: Hazard study for: ………………………..			Drg. Nos	Rev Nos
Team Members:				Date	Sheet No of Meeting No.
Hazardous Event or Situation	Caused by / sequence of events	Consequences Immediate/ Ultimate	Estimated Likelihood./ Suggested measures to reduce likelihood	Emergency Measures (reduce consequences)	Action Required

Table 3.11
Hazard study summary table

3.13 Hazard studies for computer systems

One of the difficulties facing the hazard study team for any type of controlled process is how to account for any hazards that may be introduced by unplanned actions of the control system. The problem must be faced because the requirements of phase 3 of IEC 61508 call for us to *'determine the hazards and hazardous events of the EUC and the EUC control system...'*

Where a simple stand-alone controller performs the control function it is easy to consider that the controller may fail in a limited set of modes, e.g. output to a valve could go high, low or frozen or even oscillating.

Where the control function is performed by a programmable control system such as:

- A single PLC with operator interface panel
- A distributed PLC based control system with network connections to I/O units and operator workstations

- A process control system DCS
- There is considerable potential for a whole range of malfunctions

3.13.1 Examples of potential causes of failures

- Human operator errors due to mistaken identities of controllers. (A common problem with shared display units)
- Incorrect entry of set point values, wrong valves set to manual mode
- Networked systems may be vulnerable to external corruption caused by unauthorized access within the organization or even via Internet
- Power failures within the network may lead to some control functions shutting down whilst others continue normally
- Software in large systems is frequently updated and it can be difficult to control. Small changes to configuration of the control functions are often made and may have unforeseen effects on other control functions that are sharing data

3.13.2 Guidelines

Various adaptations of hazard study techniques have been developed by end user companies to deal with the need to include the assessment of computer systems within the framework of hazard studies. The system developed by ICI in the early 1990s has been applied within AECI to a limited extent and a brief outline is given here. A similar system is described in the UK Defence Standard 00-58. Part 2 of the same standard goes into greater detail and is also available as a free download from www.defstan.mod.uk.

3.13.3 Outline of 'Chazop'

Chazop timing diagram

The diagram shows that progressive levels of Chazop study should accompany the hazard 1, 2 and 3 levels of study for the process (or EUC).

Level 1 Chazop

The first level of study covers the definition of the scope of functionality of the proposed computer system. A phase 1 checklist is used by the study team to define:
1) Fundamental role of the system
    - Protection
    - Control
    - Supervisory
    - Monitoring
    - Data logger
2) Type of System
    - Network
    - Shared display
    - DCS or PLC
    - Programmable instrument
    - Personal computer etc
3) Boundary of influence
4) Does the system play a role in site emergencies?
5) Environmental hazards such as corrosive gases: effects on system, peripherals, I/O racks, cabling

6) Sensitivity of system to power losses and surges: degree of power protection needed
7) Potential for hazardous situations arising from faults or defects in:
   - System access
   - Links within the system boundaries (networks)
   - External networking links
8) Any interactions proposed between the computer system and any safety related systems such as SIS or Fire and Gas detection systems?
9) Define the philosophy for the provision of any safety critical functions to be performed in the plant. Consider the following:
   - Software trips, interlocks and alarms
   - Hardware trips, interlocks and alarms

The outline scope of the computer system so defined can then be combined with the process flow sheets used by the hazard study team to support the identification of potential hazards at study levels 1 and 2.

3.13.4 Hazard study 3 Chazop

As the design of the plant and the control system progresses the details available from the functional specification of the control system will permit a more detailed hazard study just as is the case with the process hazard study. ICI and the AECI HAZOP manual suggest a detailed checklist procedure is then applied in a study team format to cover all the potential hazards.

For the evaluation of the effects of links across boundaries the method suggested by ICI and the defense standard is to use the typical hazard study 3 guide words (more of, less of, none of, other than, sooner/later than, corruption of, what else, reverse of) applied to data transfers between major components of the system.

Extent of study

The extent to which such techniques should be used depends on the nature of the proposed control system. If it is a prototype system built up from a range of devices it may be best to apply an exhaustive set of checks on the effects of data transfer faults. If it is a well-established DCS or proven network the study may be limited to data flows between field devices and the I/O subsystems.

Final assessment of computer system

Finally a detailed assessment procedure based on extensive checklists is proposed for the purpose of validating the design and installation of the computer system. The checklists are too long to go into in this book but are designed to ensure that all sensible precautions based on good engineering practices have been taken to reduce the potential for malfunction of the hardware and software, operator interfaces and power distribution etc. This level of study is suitable for detailed approval of the computer system but is not appropriate for the purpose of identifying potential hazards arising from the EUC control system.

3.14 Data capture checklist for the hazard study

Intended for use by instrument engineers for development of SIS performance requirements, to be used in conjunction with hazard study working sessions.

Note 1: A company risk profile chart often defines risk categories. The consequence and frequency parameters on these charts can usually be seen as range values.

Note 2: Items 11 and 12 are used where risk graphs are available for SIL determination. Items 4 and 5 will also be needed.

	Information to be obtained	Data/Description
1	Hazop record item no	
2	Equipment ref./description State number of units	
3	Hazardous event description and sequence of events leading to it.	(Or see hazop record sheet if adequate)
4	Estimated hazardous events/yr due to all causes and without protection (i.e. demand rate) or equivalent classification code if a risk profile chart is in use. See note 1.	(Risk graph parameters: W1, W2, or W3)
5	Consequence description See note 1.	(Risk graph parameters: Ca, Cb, Ce or Cd)
6	Applicable risk category codes. (if such codes have been defined by the organization)	
7	Tolerable frequency for the hazardous event at the site.	
8	Tolerable frequency per unit (divide above by no of units)	
9	Any additional non-instrument measures to reduce risk? i.e. layers of protection	
10	Any additional instrumented measures to reduce risk? E.g Alarms, DCS based interlocks?	
11	Is personnel exposure continuous or only a small time during the shift? For Enviro and Financial losses use Fa	Continuous = Fa Small fraction = Fb
12	What is the probability that plant will be able to avoid the hazard if the trip were to fail on demand? State either: High or Low	High = Pa Low = Pb
13	Basic description of the trip function. Note any operating modes that require bypass of the trip.	
14	What is the safe state of the process to be reached after trip action? What is the available time to reach the safe condition?	
15	What are the risks created by a spurious trip? What are the costs per trip?	
16	What will be an acceptable frequency for spurious trips (e.g. 1 per 10 years) per unit of equipment?	

Table 3.12
Data capture checklist

Application: The data obtained in this worksheet could be used for:

- Overall safety requirements definition as per IEC 61508 phase 4
- Allocation of risk reduction measures to SIS and non-SIS functions as per phase 5
- Determination of target SILs for the SIS, either by quantitative or by qualitative means.
- Drafting of the SIS safety requirements specification.

4

Safety requirements specifications

4.1   Developing overall safety requirements

In this chapter we are going to make sure that we know how to prepare a safety requirements specification (we shall have to call it SRS from hereon). This task is made easier for us by the fact that the ISA and IEC standards have spelt out the contents of the SRS very clearly.

Our first task is to make sure we have some continuity from the previous material on hazard studies since it is the transition from hazard study to safety requirements specification that is so critical to the quality of the SIS solutions. We shall look at the development phases where the overall safety requirements are defined and then risk reduction tasks are allocated to SIS and non-SIS contributors. The development process leads to the actual SRS for the safety system that is to be designed and installed.

Most companies will need to develop their own standard specification format for the SRS. We shall look at the component parts of the SRS leading to a checklist that could be used to provide a pro-forma specification document.

There are one or two well-tried documentary methods for defining the trip logic that should be considered.

Finally we take a look at some methods of determining SILs that may be used in the course of arriving at the SILs for each individual safety function. Various methods for determining SIL requirements have been developed in the past and most of these have now been captured into the IEC and ISA standards in the guideline sections. It's a subject that causes considerable difficulties in organizations, perhaps because it is not an exact science and there can be a lot of expense at stake. These methods depend on the quality of information flowing from the hazard studies and thus provide continuity in the safety life cycle.

4.1.1   Components of the SRS

It is generally convenient to structure the SRS into three main components as shown below:

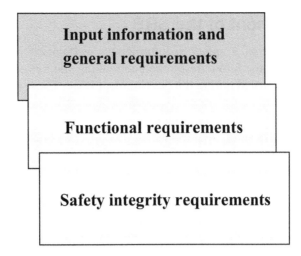

Figure 4.1
The three components of a safety requirements specification

4.1.2   SRS input section

The input information section:

- Provides information on the process and the process conditions;
- Defines any regulatory requirements;
- Notifies any common cause failure considerations;
- Lists all safety functions covered in the SRS 4.1.3 SRS functional requirements

4.1.3   SRS functional requirements

The functional requirements specification defines for each function:

- The safe state of the process
- The input conditions leading to response
- The logic
- The output actions to bring the process to a safe state
- The process safety time or speed of response

4.1.4   SRS integrity requirements

The integrity requirements specification defines for each function:

- The required risk reduction and the SIL
- Expected demand rate and the low or high demand mode
- Energize or de-energize to trip
- Spurious trip constraints
- Proof testing strategy
- Environmental conditions

4.2   Development of the SRS

The objective of a safety requirements specification is to capture all the information necessary for the future design and continuing support of a safety function. Two versions of SRS are created during the development stage. These are:

- Overall safety requirements
- SIS safety requirements

In IEC 61508, Phase 4 defines an overall safety requirements specification. This describes the risk reduction requirements for overall safety including the non-SIS and external protection functions. The safety allocations phase (5) then allocates risk reduction duties to both SIS and non-SIS protection layers for each safety function.

The SIS safety requirements are then defined at the start of phase 9 on the basis of the known requirements for all risk reduction contributors. The structure of the two levels of SRS is the same and we can start by looking at the overall SRS and then adding in more specific details for the SIS.

4.2.1   General development procedure

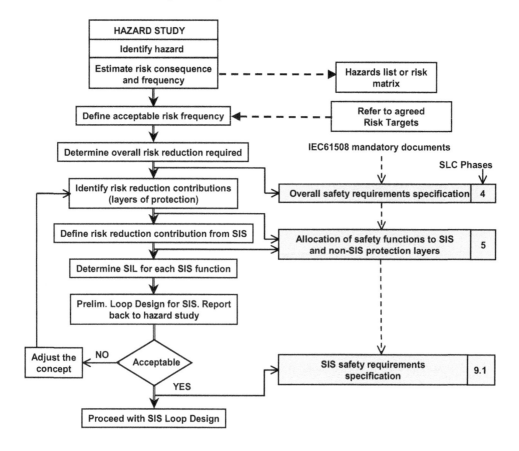

Figure 4.2
General procedure for developing and specifying safety requirements

Fig 4.2 describes the activities involved in progressing from the hazard study stage through to the completion of the detailed safety requirements specification. The development of the SRS is an iterative process carried out by the instrument engineer in co-operation with the plant design team and any associated safety specialists. Most of the information required for the SRS flows from the hazard analysis stages as we have seen in Chapter 3.

The SIS safety requirements specification should be developed in association with the non-SIS protection layers. The IEC 61508 standard calls for the overall safety requirement to be defined first in phase 4 followed by an allocation phase 5, which defines the sharing of protection duties across the layers of protection. The final SRS for the safety-instrumented system is then part of the detail design activities for the SIS known in IEC terms as the 'realization' phase (phase 9). This procedure can be rather confusing at first but it appears to be designed to ensure that the basics of the SRS are in place and verified before the design team goes too far with the technical specifications for the SIS.

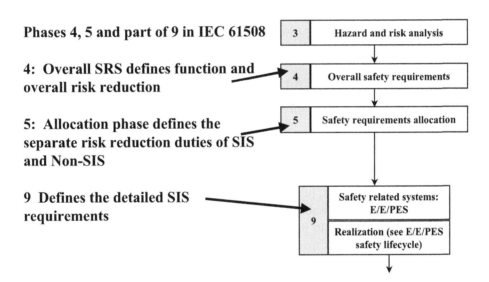

Figure 4.3
Safety life cycle phases leading to the detailed safety requirements

Fig 4.3 shows how the SRS and allocation phases are positioned in the IEC safety life cycle model. It is practical to execute these two phases together as part of the iterative design process whereby trial solutions are evaluated and adjusted until the best balance between SIS and non-SIS protection is achieved. Fig 4.4 shows the iterative process whereby trial solutions are reviewed, often with the aid of additional hazard analysis work until the overall combination of protection measures is considered to be both feasible and satisfactory.

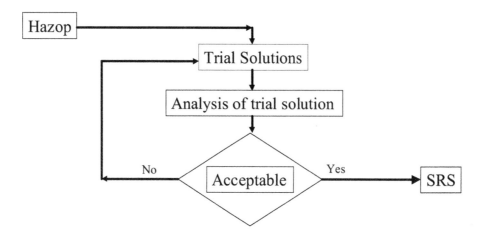

Figure 4.4
The iterative procedure to arrive at an acceptable safety solution

Phase 4 and 5 often involve looking ahead to the conceptual design stage since we need to have a good idea of what is achievable when the SRS is being firmed up.

So let's now look at expanding the contents list for the SRS. This will lead to a basic checklist for compiling a complete specification.

4.2.2   The input requirements

The development work begins with assembling basic data from the preceding stages of the SLC.

Overall SRS input requirements

- SRS subject, title, ref number, etc.
- Process data sheets, chemical hazard data sheets.
- References to process descriptions, flow sheets or P & IDs correctly defining the scope of the plant.
- Regulatory requirements applicable to the plant or the process.
- List of safety functions to be included in the SRS.
- Common cause failure possibilities.

4.2.3   Developing the functional requirements

The first thing we need to be clear about is that each individual safety function has to be specified. This means that there has to be a separate specification sheet for each function. Possibly the best way to deal with this is to have a safety functions list in the common input section of the SRS, with a data sheet for each function.

Functional requirements for the overall SRS

IEC 61508 calls for the following:

For the overall SRS: 'the safety functions necessary to ensure the required functional safety for each determined hazard shall be specified'.

This phase asks for the overall risk reduction requirements for each function but does not require technology specific solutions at this stage. However it does have some very significant things to say about the basis of the specified risk reduction as summarized below.

Overall SRS contents as required by IEC 61508

- All risk reduction measures to be specified.
- EUC control system fail to danger rate to be justified.
- EUC control system to be separate from SIS.
- Alternative: EUC control system designated as SIS.

We should take particular note of the following requirements:

The basic (EUC) control system very often plays a part in the potential event rate for a given hazard. Hence the fail to danger rate claimed for the basic control system is part of the risk assessment. At paragraph 7.5.2.4 the IEC standard calls for the dangerous failure rate claimed for the EUC control system to be supported by data or experience. It sets a minimum failure rate of 10^{-5} failures per hour (which is approximately 1 dangerous failure per 10 years). All foreseeable dangerous failure modes are to be taken into account. In practice this means that the complete instrument loop or set of controls cannot be credited with a performance better than 1 failure in 10 years and for whatever performance is claimed the designers are expected to justify the figures.

EUC control system may have to be a safety system: If the EUC is to play a role in safety as part of the risk reduction measures it cannot be assigned a SIL level of 1 or higher unless it is treated as a safety related system. This is the problem we looked at in Chapter 2.

Functional requirements

Allocations phase

For the safety allocations phase the designated safety related systems that are to be used to achieve the required functional safety shall be specified. The necessary risk reduction may be achieved by

- External risk reduction facilities
- E/E/PES safety related systems
- Other technology related systems

This phase allocates proportions of risk reduction for each function to independent protection layers. Hence the SIS duties can be seen in context of the other protection systems. Clearly if the other protection systems are changed at a later stage the duties of the SIS may also have to change. By this method the IEC standard tries to ensure that overall safety is not compromised through modifications to either SIS or non-SIS equipment.

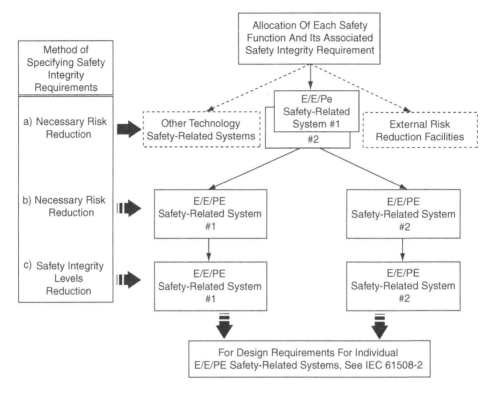

Figure 4.5
Allocation phase

The information needed to complete the allocation phase concentrates on the safety integrity aspects of the risk reduction systems. We shall look at these in the next section. The functional requirements for the SIS should at least be outlined at this stage.

Once a realistic and achievable allocation of risk reduction measures has been agreed the development of details for the SIS safety requirements can proceed within a framework that is provided in phase 9 of the safety life cycle.

Functional requirements for the realization phase

To complete the detailed SRS all we need to do is to develop the engineering information to satisfy the list of requirements provided in the standards.

The following list of mandatory requirements for the SIS specification are taken from ISA S4.01 clause 5.3

- Define safe state of process for each event
- Process inputs and trip points
- Normal operating range
- Process outputs and actions
- Logic solver functions/relationships
- De-energize or energize to trip
- Manual shutdown considerations
- Action on loss of energy to SIS
- Response time to reach safe state
- Response action for overt fault
- Human–machine interfaces
- Reset functions.

All of the above items are found in the IEC standard in phase 9 where more detailed explanations of these terms will be found. Details of phase 9 are set out in part 2 of IEC 61508.

4.2.4   Safety integrity requirements

In the overall SRS the safety integrity is simply the required overall risk reduction for each listed function.

In the allocation phase the safety integrity is the risk reduction allocated to each of the SIS and non-SIS contributors. There are some noteworthy points in the IEC standard at this phase:

- The IEC standard requires that we take into account the skills and resources available at a particular location when working out a suitable safety integrity value.
- In some applications a simpler technology solution may be preferable to avoid complexity.
- Iterative design may be needed to modify the proposed architecture of the SIS to ensure it can meet the required SIL.
- For each function state if the target safety integrity parameter is to be based on high or low demand mode of operation.
- Allocation to use techniques for combination of probabilities. This includes quantitative and qualitative methods for determination of SILs.
- Allocation to take into account the possibility of common cause failures between the protection functions.
- As the allocation proceeds the SIL values of the SIS for each function are to be listed; basis of high or low demand to be stated.
- Shared equipment to be specified to the highest SIL function.
- No single SIS to have an SIL of 4 except by special demonstration/experience.
- No SIS to have an SIL of less than 1.
- All information to be documented.

Using the above requirements the completed allocation phase will provide a well-documented baseline for the completion of the detailed safety requirements specification for the SIS. The safety integrity requirements for the SIS are then developed to conform to the following requirements summarized from phase 9.

Safety integrity for the detailed SRS

For each safety function:

- The required SIL and also the target risk reduction factor where quantitative methods have been used
- High or low demand mode
- Requirements and constraints to enable proof testing of the hardware to be done
- Environmental extremes
- Electromagnetic immunity limits needed to achieve EMC

4.2.5   Conclusions on the SRS development

We have seen how the safety requirement specification has been expanded from overall safety requirements through allocation into a detailed SRS for the safety instrumented system. The result of these activities will be a fully documented record of the design requirements and the history of where the requirements have come from.

We are now in a position to consider some documentation aids for the SRS.

4.3   Documenting the SRS

What tools can be used to capture the requirements efficiently? The main needs are:

- A specification sheet blank form listing all the technical data items required for completing a safety requirements specification.
- A means of representing the functional requirements that is accurate but flexible enough to accommodate changes.
- A systematic means of determining the target SILs.

Tools for the first two items are suggested here. Methods for SILs are considered in the next section.

4.3.1   Checklist for SRS

On the basis of the preceding notes we can now set out a checklist or pro-forma SRS document plan as shown in the following tables. The normal practice in an operating or engineering company will be to develop a company specific SRS template for use in all projects. The template will carry header sections defining the exact reference numbers and descriptions for the particular safety function. There will be a complete document for each individual safety function although it is possible that the general section will be common to a set of applications. In this case the general section will list each of the safety functions covered by the document.

General Section	
Title and subject of the SRS	
Spec. no/Date/rev no/Author	
P & I Ds	
A list of the safety functions required	
Cause and effect matrix	
Logic diagrams	
Process data sheets	
Hazard study report references	
Process information (incident cause, dynamics, final elements, etc) of each potential hazardous event that requires an SIS	
Process common cause failure considerations such as corrosion, plugging, coating, etc	
Regulatory requirements impacting the SIS	
Spec to be derived from the allocation of safety requirements (IEC phase 5) and from safety planning. Doc Ref to be given	

Table 4.1
General requirements checklist for SRS

Safety Functional Requirements	
Details sufficient for the design and development of the SIS	
Is control to be continuous or not	
Is the demand mode low or high	
The definition of the safe state of the process, for each of the identified events	
Method of achieving safe states	
Interfaces with other systems	
All relevant modes of operation	
Significance of all hardware/software interactions…constraints to be identified	
SIS start up/restart requirements	
The process inputs to the SIS and their trip points	
The normal operating range of the process variables and their operating limits	
The process outputs from the SIS and their actions	
The functional relationship between process inputs and outputs, including logic, math functions, and any required permissive logic	
Selection of de-energized to trip or energized to trip.	
Consideration for manual shutdown.	
Action(s) to be taken on loss of energy source(s) to the SIS	
Response time requirements for the SIS to bring the process to a safe state	
Response action to any overt fault	
Human–machine interfaces requirements	
Reset functions	

Table 4.2
Safety functional requirements checklist for SRS

In IEC 61508 part 2 you will find detailed descriptions of the points given in our checklists for the specification of the SIS (referred to in the standard as E/E/PE safety related system). These are used in the first part of phase 9 of the safety life cycle and are to be found in paragraph 7.2.3.

Safety Integrity Requirements	
The required SIL for each safety function	
Low or high demand mode for each function	
Environmental extremes	
Electro magnetic immunity limits for achieving EMC	
Requirements for diagnostics to achieve the required SIL	
Requirements for maintenance and testing to achieve the required SIL	
Reliability requirements if spurious trips may be hazardous	
Techniques to avoid specification errors. Use of IEC 61508-2 table B1. See note 1	

Table 4.3
Safety integrity requirements checklist for SRS

Note 1: The IEC standard requires that the designer should use 'an appropriate group of techniques and measures' to avoid errors in the specification stage. Table *B1* and annex *B* of IEC 61508 part 2 describe techniques for avoiding systematic errors in specification work. This an important point to note because an incorrect assumption built into the SRS may carry through into the build and test phases and become a permanent 'systematic error' in the design.

4.3.2 Defining the functions

We need a reliable method of defining the functional relationships between inputs and outputs in the SIS. Usually we have to circulate the information among many parties and it has to be available at the plant for use in operations, training and in future studies. Whilst many companies will define their requirements by a 'narrative description' there are benefits to be found in supporting the narrative with graphical description methods. Hopefully these methods will avoid the risk of misunderstandings between designer and builder.

Important considerations for choosing diagram formats are:

- They must be easy to work with at the development stage where frequent redrafting is involved.
- They should be suitable for creating in electronic format for use on any PC and for transmission of files to other parties.
- They must ensure that strict control of revisions is simple to achieve and that change histories can be clearly shown.
- There must be no chance of misunderstanding or ambiguity.

There are various established ways of graphically presenting the information and we are going to see three examples. Let's take the example of a proposed safety function for a reactor protection task as outlined in Figure 4.6 and begin with a narrative description of the functional safety requirements.

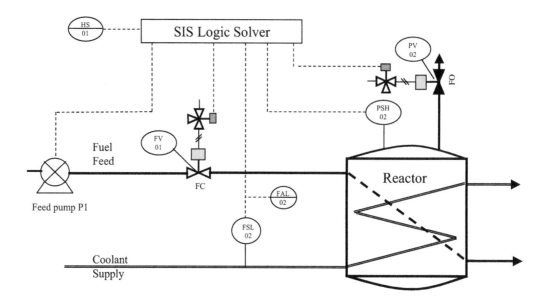

Figure 4.6
Proposed trip system for a reactor

Narrative description

The safety function is required to protect the reactor against overpressure or events that could lead to overpressure. The function subdivides into three possible event responses or sub-functions.

Sub-function 1: Loss of coolant. The process feed to the reactor must be shut off automatically if there is a loss of coolant detected by the logic solver reading of flow meter FT-02. The feed is to be shut off by releasing the air supply from FV-01 and this is to be followed 30 seconds later by tripping of the feed pump P1. The reactor vent valve PV-02 is to be opened two minutes after tripping FV-01.

Sub-function 2: High pressure event: If the pressure switch PSH-02 detects high pressure in the reactor the vent valve PV-02 is to be opened immediately and the process feed is to be shut off as per function 1.

Sub-function 3: Manual shutdown or loss of services: Sub-function 1 and Sub-function 2 will also be initiated by operating the manual shutdown pushbutton HS-01. In the event of loss of instrument power or loss of air supply trip valves FV-01 and PV-02 shall be released immediately and the pump P1 is to be stopped.

Matrix or cause and effect diagram

The above narrative description can be stated in the form of an input/output table or cause-and-effect matrix as shown in Figure 4.7.

The diagram shows the states of inputs on the left side and the states of outputs on the top right. The required output states corresponding to input states are defined by crosses at the intersections. The diagram assumes that all initiator signals are arranged to hold the tripped condition in the logic solver until reset by operator action. This method of defining functions is good for placing on P&I drawings and for explicitly defining interlocks and basic trip logic. It is not so good for defining shutdown sequences. Note how the SIL requirement for each sub-function has been placed on the diagram. This is an option that may be of value but should not conflict with the safety integrity requirements section of the SRS.

Instrument tag	Description	Instrument range	Units	Trip setting	Open valve PV02	Close valve FV01	Stop pump P1	Start timer T1	Start timer T2	Required SIL
FT-01	Coolant flow low (FSL)	0-100	m³/h	15		X		X	X	2
PSH-02	Reactor pressure high	0-2000	kPag	1600	X	X		X	X	2
HS-01	Manual trip				X	X	X			2
XS-03	Loss of power				X	X	X			2
PSL-04	Loss of instrument air	0-1000	kPag	300	X	X	X			2
KY-01	Timer T1 expired	0-300	secs	30			X			2
KY-02	Timer T2 expired	0-300	secs	120	X					2
	Note: All initiators are to stay tripped until reset by operator.									

Figure 4.7
Cause-and-effect diagram or matrix table

Trip logic diagrams

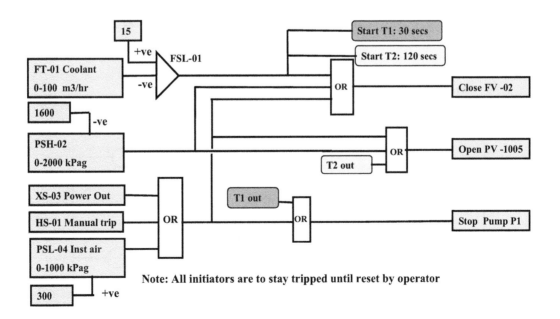

Figure 4.8
Trip logic diagram

Trip logic diagrams are widely used to define safety functions. They are useful because a complete scheme can be followed on one drawing and also because many graphical programming systems use a similar format. Figure 4.8 shows how the logic seen in the matrix table will look in this form. Note that in this version a 'true' or '1' state defines the safety function action to be taken. Some versions are drawn to show the '1' state as enabling the process to run and the trip is defined by '0' states. The latter will indicate fail-safe states of the equipment but can be more difficult to follow. The objective is to define the safety function requirements as clearly as possible and leave the fail-safe design to the next phase of the SLC.

Trip link up diagrams

Trip link up diagrams allow a complex trip function to be represented on a multiple set of engineering drawings. This type of diagram is very useful when used as a means of defining the overall scheme to a design contractor because the task can be broken down into easily identified small sections. It also provides a convenient format for adding descriptions and data.

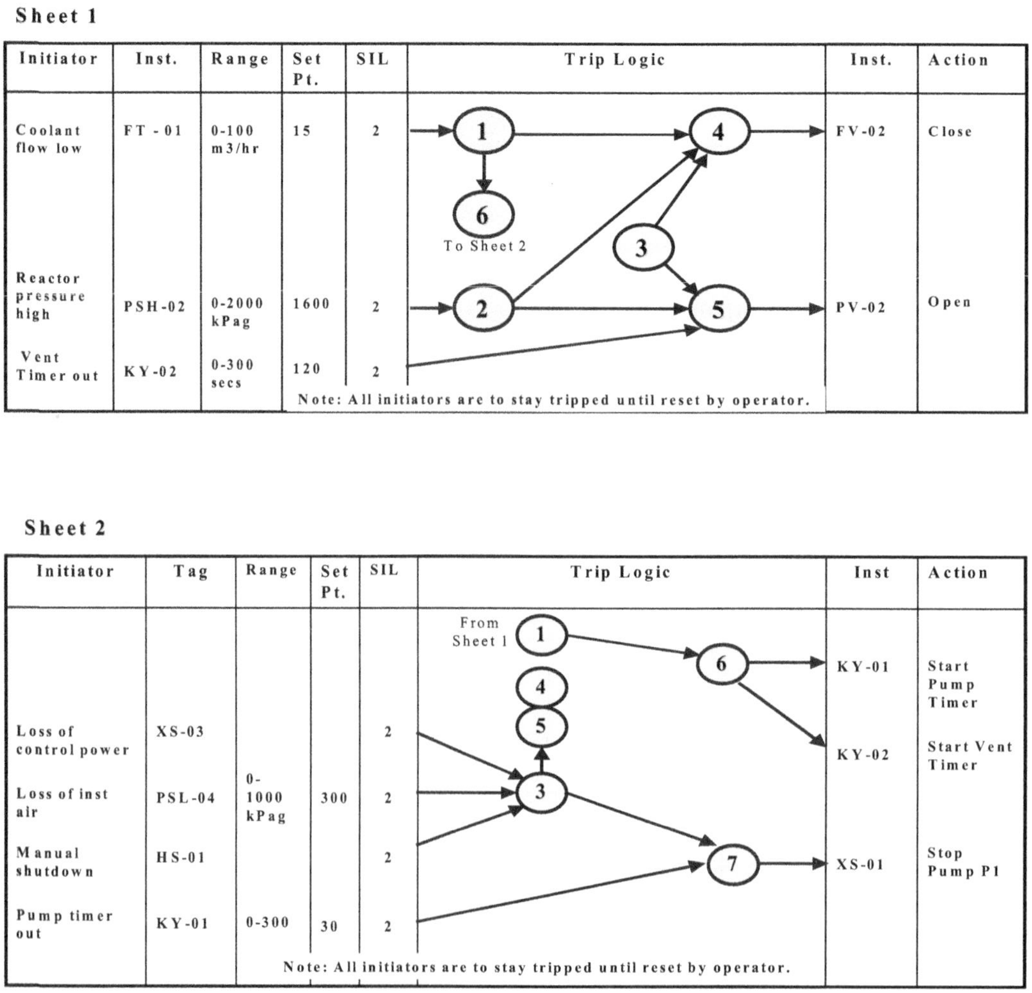

Sheet 1

Sheet 2

Figure 4.9
Trip link up diagram

Note of caution: It is attractive to consider using graphical software tools for defining the safety function and it can be even more efficient if the graphical tools can be used to directly generate the application program for a safety controller system. However we must bear in mind that the SRS is a specification document and should be used to ensure the original functional requirements are defined independently of the final implementation.

There is a potential risk that by defining the safety function directly in the programming tool any errors in the original specification will be automatically copied through to the implementation without an intervening stage of verification.

4.4 Determining the safety integrity

As we have seen one of the most important tasks in the SRS development is to specify the safety integrity of each SIS function. This needs to be done fairly early in the development stages to see that our proposed solutions are realistic, achievable and of course affordable. The cost of the SIS will rise steeply with the SIL values even if we buy a logic solver that meets SIL 3 the cost of sensors and actuators and engineering work will still be influenced strongly by the SIL rating.

It is important therefore that we have a consistent method of arriving at SIL values within any given organization. To do this we need to consider the most appropriate method of determining the SIL for any particular function. There are at least 3 recognized methods of doing this and these have been widely documented over recent years and now are built into the ISA and IEC standards.

4.4.1 Diversity in SIL methods

The reason for such diversity in methods of determining SILs is probably due to the difficulties of arriving at reliable and credible estimates of risk in the wide variety of situations faced in industries. Whilst a quantitative risk assessment is desirable it may be worthless if the available data on fault rates is minimal or subject to huge tolerances. Qualitative methods allow persons to use an element of judgment and experience in the assessment of risk without having to come up with numerical values that are difficult to justify.

One advantage of the SIL concept is that it provides a 10:1 performance band for risk reduction and for SIS in each safety integrity level. Hence the classification of the safety system can be matched to a broad classification of the risk and the whole scheme is able to accept a reasonable tolerance band for the estimates of risks and risk reduction targets.

4.4.2 Summary of methods for determination of SILs

We have already seen that the methods divide into quantitative and qualitative types. IEC 61508 part 5 outlines one qualitative method and 2 quantitative methods, namely:

- Quantitative method using target risk reduction factors.
- Qualitative method using risk graphs.
- Qualitative method using hazardous event severity matrix.

For the process industry sector the newly released standard IEC 61511 provides more specific details of the established methods. These are set out in IEC 61511 part 3 and consist of:

- Quantitative method using target risk reduction factors.

- Qualitative method using risk graphs, variations are shown for use where the consequences are environmental damage or asset loss.
- Qualitative method using safety layer matrix.
- Qualitative/quantitative method using layers of protection analysis (LOPA).

The last two items above are very similar in nature and use formalized definitions of safety layers and protection layers to allow hazard study teams to allocate risk reduction factors to each qualifying layer of protection.

We take a brief look here at the 3 methods for SIL determination described by IEC 61508. You are advised that before attempting to carry out an SIL determination within an project you should carefully read through the material in IEC 61508 part 5 or study IEC 61511 part 3.

4.4.3 Quantitative method

We have already been introduced to this procedure in Chapter 3 but a further review using a simple application example may be helpful at this point. We use fault tree analysis and diagrams as introduced in Chapter 3. Firstly recall that risk reduction comes in parcels! The IEC risk reduction diagram shown here applies.

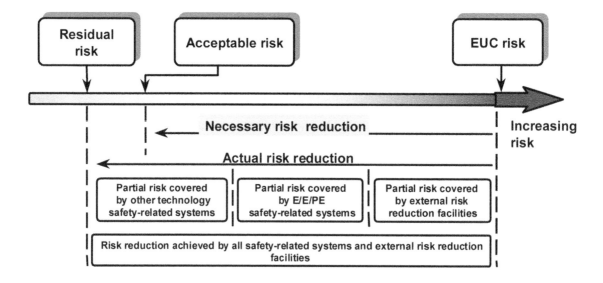

Figure 4.10
IEC risk reduction diagram

4.4.4 Design example

Adding protection: Let's return to the simple high level hazard example shown in Chapter 2.

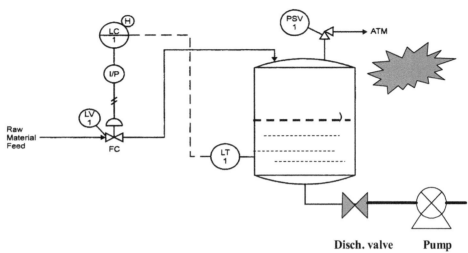

Basic tank level control with over pressure release hazard

Figure 4.11
High level hazard

Firstly we can draw a fault tree identifying the cause of the hazard and estimating the consequences as a possible fatality with estimated fatal accident rate.

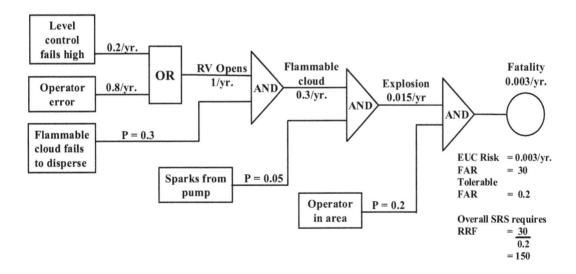

Figure 4.12
Fault tree for the unprotected hazard

We can set the tolerable risk frequency either by using the classification table we saw in Chapter 2 or by setting the FAR target. In this case we have set FAR target at 2.

External or mitigation layer

Now we can add an external means of risk reduction in the form of a fence around the offending vessel so that the probability of an operator being nearby when the explosion occurs is substantially reduced. Note that this action does not change the hazard event rate but it does reduce the risk frequency of a fatality.

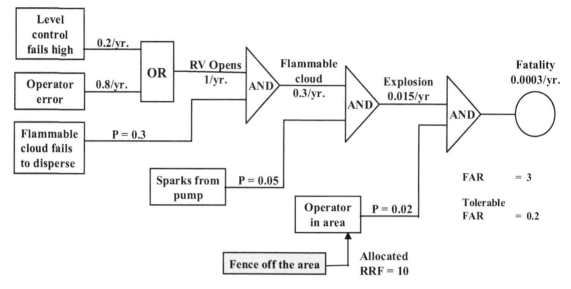

Figure 4.13
Fault tree with external protection layer

In this case the FAR remains well above target and this leads us to suggest an SIS protection layer.

Adding the SIS

The suggested SIS is the high-level trip system designed to reduce the risk of overfilling and over pressurizing in the tank.

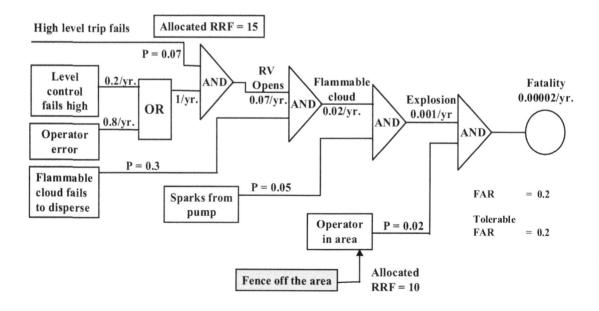

Figure 4.14
Fault tree with SIS and external protection layer

We can calculate the desired RRF for the SIS using this fault tree. Alternatively we can draw a risk reduction model or diagram as shown in the next diagram.

Risk Reduction Model

Showing safety integrity allocations for tank example

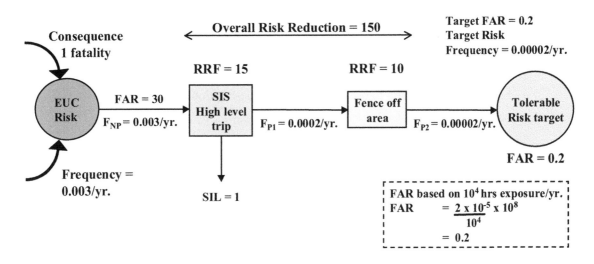

Figure 4.15
Risk reduction model for the protected hazard

Note how the risk reduction factor can be calculated for each component of the safety system. Note how the failure rate target figures can be adjusted within a feasible range until a realistic model of the protection system is achieved.

Obtaining the SIL

Now that we have isolated the risk reduction factors we can see how the safety integrity level can simply be obtained from the tables in the standards just as we did before in Chapter 2.

This exercise results in requirement for a safety integrity level of 1 to be met by the SIS. Assuming experience confirms that this is reasonable and feasible for the required task the result would be acceptable and the requirement would be confirmed into the SRS.

4.4.5   Summary of quantitative method

What we have seen here is an example of the quantitative method being used to assist in the development of the SRS and the defining of the SIL.

In summary:

1. Evaluate hazard event rate without protection. Define target risk frequency. Record all details under phase 4 of the SLC.
2. Add external and non-SIS protection and evaluate effect on risk frequency.
3. Propose an SIS risk reduction measure which reduces the hazard event rate and hence the risk frequency.

4. Conclude a practical risk reduction factor for the SIS consistent with being below the target risk frequency.

5. Convert the risk reduction factor to an SIL value for the SIS.

6. Draft the SRS with a reference to the calculation sheet and risk reduction model. Record all documents into phase 5 of the SLC.

7. Finalize SIS detail SRS as part of phase 9 of the SLC.

4.4.6   Risk graph methods

The origin of this method is the German standard VDE 19250 which arrived at an SIL classification code of AK-1 to AK-8.

This method is a very attractive alternative for arriving at SILs because it avoids the need to place actual quantitative figures on the hazard demand rates, risk frequency and the consequences. Since in many cases the figures we use are very approximate it is perhaps more realistic to use an approximate description.

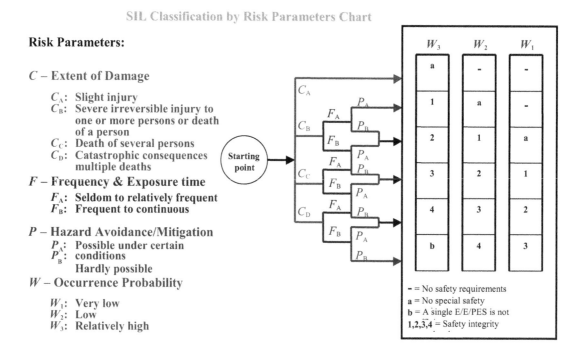

Figures 4.16
Risk parameters chart based on IEC 61508 example

You can see how it works by looking at the next diagram. Let us test it for our previous example.

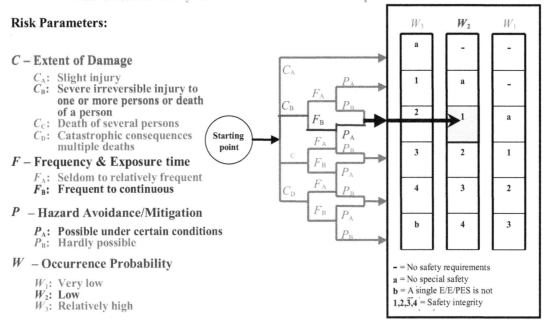

SIL Classification by Risk Parameters Chart: Example

Risk Parameters:

C – Extent of Damage

C_A: Slight injury
C_B: Severe irreversible injury to one or more persons or death of a person
C_C: Death of several persons
C_D: Catastrophic consequences multiple deaths

F – Frequency & Exposure time

F_A: Seldom to relatively frequent
F_B: Frequent to continuous

P – Hazard Avoidance/Mitigation

P_A: Possible under certain conditions
P_B: Hardly possible

W – Occurrence Probability

W_1: Very low
W_2: Low
W_3: Relatively high

- = No safety requirements
a = No special safety
b = A single E/E/PES is not
1,2,3,4 = Safety integrity

Figure 4.17
Example of decision path in the risk parameters chart

For the unprotected plant:
Consequences: *Cb*
Frequency of exposure or exposure time in zone: *Fb*
Possibility of avoidance: *Pa*
Probability: *W3*

Which leads us to an overall solution requirement of SIL 2.

If we then add the non-SIS measures such as the fence we can test again for the task to be performed by the SIS.

In this case F_b changes to F_a and the SIL becomes SIL 1. So the risk graph enables us to decide on the SIL based on our assessment of the risk when there is no SIS but it does allow us to include for the presence of other protection measures.

It seems a lot easier to arrive at the same conclusion that we achieved with the quantitative method. However, the accuracy depends on the interpretation of the clauses.

4.4.7 Defining parameters and extending the risk graph scope

Before a risk graph can be used the project team must establish the definition of the parameters being used and decide on the design of risk graph to be used. In practice in the process industries there are separate versions for three categories of hazard i.e.:

- Harm to persons
- Harm to environment
- Loss of assets (production and equipment losses/repair costs)

All three versions of the risk graph can have the same basic layout but for environment and asset loss the parameter *F*, for exposure, is considered to be permanent and can be left out of the diagram. For a full determination of SIL requirements each safety function should be evaluated for the three categories of hazard and the SIL target rating must be set to meet the highest value found from the three categories.

4.4.8   Risk graph guidance from IEC 61511

IEC 61511 has generated a very useful example version of the factors affecting the parameters C, F, P and W. We must be clear that for each application it is the responsibility of individual companies or safety departments to establish their own agreed parameters for the risk graph they wish to use. The meaning of the parameters must first be clear. Table B1 from IEC 61511, seen below, can be used for this. In particular it is important to note the interpretation of the term W as being based on the assumption that no SIS is present.

Parameter descriptions table

Parameter		Description
Consequence	C	Average number of fatalities likely to result from the hazard. Determined by calculating the average numbers in the exposed area when the area is occupied taking into account the vulnerability to the hazardous event
Occupancy	F	Probability that the exposed area is occupied. Determined by calculating the fraction of time the area is occupied
Probability of avoiding the hazard	P	The probability that exposed persons are able to avoid the hazard if the protection system fails on demand. This depends on there being independent methods of alerting the exposed persons to the hazard and manual methods of preventing the hazard or methods of escape
Demand rate	W	The number of times per year that the hazardous event would occur if no SIS was fitted. This can be determined by considering all failures which can lead to one hazard and estimating the overall rate of occurrence

Table 4.4

Deciding parameter values in a risk graph

Risk tables must align with any existing risk profile specified by a user company for its operations and sites. IEC 61511 contains a table of suggested consequence and frequency parameters that might be used to produce what is described as a 'semi-quantitative risk graph'. Other authorities such as UK Offshore Operators Association have produced their own consensus parameter descriptors. Hence various well-established risk graph versions can be used in relevant industries.

The values for personal hazards suggested by IEC 61511 part 3 Annex D are indicated in the following table, but please note that the IEC standard and Annex D should be carefully studied for the exact wording and context. Our notes here are intended for preliminary guidance only:

Parameter	Range of values
Consequence: C = Average number of fatalities	C_A = Minor injury
	C_B = Range 0.01 to < 0.1
	C_C = Range 0.1 to < 1.0
	C_D = Range > 1.0
Occupancy (F) This is calculated by determining the length of time the area exposed to the hazard is occupied during a normal working period	F_A = Rare to more often exposure in the hazardous zone. Occupancy less than 0.1
	F_B = Frequent to permanent exposure in the hazardous zone.
Avoidance (P) Probability of avoiding the hazardous event if the protection system fails to operate.	P_A = Possible to avoid. Should only be selected if all the following are true: – facilities are provided to alert the operator that the SIS has failed – independent facilities are provided to shut down such that the hazard can be avoided or which enable all persons to escape to a safe area – the time between the operator being alerted and a hazardous event occurring exceeds 1 hour.
	P_B = Not possible to avoid. Applies if any of P_A conditions are not met
Demand rate (W). The number of times per year that the hazardous event would occur in the absence of the SIS under consideration. See note 1.	W 1 = Demand rate less than 0,1 D per year. W 2 = Demand rate between 0,1 D and D per year. W 1 = Demand rate between D and 10D per year.
Note 1: In the demand rate table, D is a calibration factor decided by the user to ensure that the residual risk achieved by using the risk graph is acceptable to the organization. It is possible that the IEC committee has introduced this term to encourage users to calibrate their risk graph designs against tolerable risk levels for their particular industry. Draft versions of this standard carried defined values for W but there has been a temptation for users to take the value as a guide to acceptable risk, which was not the intention of the table. By relating the risk graph decisions to risk reduction factors provided by SILs and comparing them with typical tolerable risk frequencies it appears that D might be typically in the range 0.1 to 0.5. However, we stress that this is for the end user to decide in terms of corporate policy and local conditions.	

Table 4.5
Table of risk parameters based on IEC 61511-3 Annex D

For an individual company the values would be adjusted to align with any existing risk matrices or risk profile charts.

IEC 61511-3 Annex D also provides guidance for parameters relevant to using the risk graph for environmental hazards and for asset loss. These scales are also likely to be modified by the end user but for the initial guidance we recommend users consult the IEC 61511 standard.

4.4.9 Calibration of the risk graph

IEC 61511 requires that the parameters table and the use of the risk graph be validated by first testing its results against examples where the SIL rating is agreed with some confidence. This can be done conveniently by comparing the results of simple cases where a quantitative risk reduction model has been drawn. In practice several practical examples would serve to confirm the calibration of the risk graph.

Another approach is to test the risk graph on well-established safety systems within existing installations where the level of risk, consequence and effective SIL values are already known.

4.4.10 Software tools using risk graphs

In applications where a large number of safety functions are to be installed it is essential to set up a systematic method of specifying and recording the reasoning and decision making involved in arriving at the SIL target for each safety function. Risk graph methods are adaptable to database management applications and provide an easy to use graphical interface for a design team to work with. If the agreed and approved risk parameter descriptions and ranges are built into the application package this will ensure that all SIL decisions are made on a consistent basis

The advantages of using a database package to record SIL decisions include the ability to maintain a life cycle record of the hazard study results and operating data that may have been used in the initial decision. The software can keep a record of all changes affecting the safety functions and is easily revisited for periodic safety audits.

A number of companies have developed software tools for SIL determination and some references have been included in Appendix A.

4.4.11 The safety layer matrix method for SIL determination

Another qualitative method described by IEC standards is called the safety layer matrix method. This is described in Annex E of IEC 61508 part 5. The same procedure is detailed in the ISA standard S84.01 Annex A.3.1, where it is called the safety layer matrix and the same principles have been included in the recently issued IEC 61511-3 in annex along with the risk graph. The origin of the method is attributed to the following well-established reference book:

Guidelines for Safe Automation of Chemical Processes, American Institute of Chemical Engineers, CCPS, 345 East 47th Street, New York, NY 10017, 1993, ISBN 0-8169-0554-1.

IEC state some basic requirements for safety layers before the logic of the matrix diagram can be used:

- Independent SIS and non-SIS risk reduction facilities.
- Each risk reduction facility is to be an independent protection layer.

- Each protection layer reduces the SIL by 1 (i.e. it must be shown to be capable of an RRF of at least 10).
- Only one SIS is used.

The method then determines the SIL for the SIS by applying the situation to a severity matrix chart such as the one shown in the next diagram.

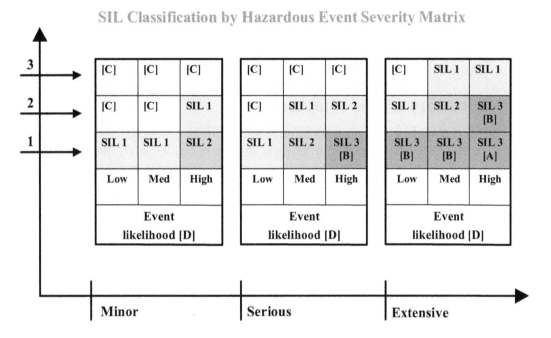

Figure 4.18
Hazardous event severity matrix

Testing this using our previous example:
- Severity: Serious event likelihood:
- Medium number of independent protection layers: 2
- SIL of the SIS = 1

This seems even easier than the risk graph but it depends on a calibrated scale of severity and the correct identification of valid protection layers.

Obviously we need to be sure that each safety layer has a suitable integrity to qualify as a protection layer. For example an alarm system would help with our example but we would have to analyze its effectiveness to be sure of its integrity.

There are some other variations on these methods detailed in the ISA annex and it is important to note that some industries, in particular the nuclear power industry prefer to determine the SIL requirements on the basis of 'consequences only'. This approach is a conservative one and probably ensures adequate protection but it could result in relatively expensive solutions.

4.4.12 The LOPA method for SIL determination

The term LOPA is an abbreviation for layer of protection analysis. The method is a continuation of the HAZOP study and accounts for each identified hazard by documenting the initiating cause and the protection layers that prevent or mitigate the hazard. The total amount of risk reduction recognized in the plant design and its protection systems is evaluated, typically using a table or spreadsheet format.

If the predicted risk level is higher than the acceptable target for the project or the company the need for additional protection is clear. If the existing risk reduction layers are given quantitative failure-on-demand values (i.e. PFD values) the additional PFD required from a safety system can be found and hence the SIL is determined from the standard SIL/PFD tables.

The version of LOPA described in IEC 61511 part 3 Annex F is interesting and very useful because it includes suggested or typical PFD values for factors such as operator responses and alarm system integrities. It also suggests a range of frequency values for demand categories. This method is effectively a general-purpose implementation of the quantitative method we have examined earlier in this chapter and previously in Chapter 2.

4.5   Summary of this chapter

This chapter has outlined what information has to be captured in the safety requirements specification and what has to be done to determine the safety integrity requirements. Three development stages are involved in moving from an overall safety requirement to a detailed performance specification for the SIS. Each stage allows the design to be placed in the context of the original protection needs so that continuity with the hazard analysis stage is maintained. Some well-established methods for determining the SILs have been described and the selection of one or more methods is a project choice dependent on company practices and the type of information available.

It is important that the project team recognizes the benefits of establishing high quality definitions for functional requirements and the benefits of optimizing the SIL requirements. It will become clear from the next chapters that the cost of safety system rises steeply with increasing SIL values; hence there is considerable value in getting the SIL target right.

If the development data has been captured properly the SLC project is now ready for the conceptual design phase.

5

Technology choices and the conceptual design stage

5.1   Introduction

The next stage in the safety life cycle brings us to what the ISA standard calls 'conceptual design'. This is all about getting the concepts right for the specific application. It also means choosing the right type of equipment for the job; not the particular vendor but at least the right architecture for the logic solver system and the right arrangement of sensors and actuators to give the quality of system required by the SRS. Here's what we are going to do in this chapter to cover the subject.

- Check the guidelines as per ISA/IEC
- Establish key design requirements
- Examine logic solver architectures; from relays to TMR
- Comment on certification

5.1.1   What does the conceptual design stage mean?

This is the stage where the control engineer prepares the whole SIS scheme from sensors through logic solver to the final element, control valve or motor trip etc. Some typical issues to be decided at this stage include;

- The decisions are made on what type of sensor system is required.
- What functions, if any, require redundant measuring sensors? What measures are needed to avoid spurious trips?
- If an instrument is prone to problems, say an oxygen detector in a gas line, this is the point where a 2 out of 3 voting (2oo3) scheme is proposed.

The selection of the logic solver technology is made...e.g. relays or PES. The architecture of the PES has to be decided. Do we need dual redundant architectures or will a single channel PES be acceptable?

What type of final element tripping device can we use. Is it serviceable and testable?

All basic design decisions are taken at this stage for the SIS but are subject to evaluation, review and finally verification, before the design is 'cast in stone' and the detail engineering proceeds. As usual in engineering projects if the right decisions can be made up front whilst the choices are open the rest of the job will go much better.

5.2 What the standards say?

5.2.1 ISA conceptual design stage

Figure 5.1 shows us where we are in the ISA life cycle model and points us to paragraphs 4.2.7 and clause 6 of the standard where the ground rules for the conceptual design stage are set out.

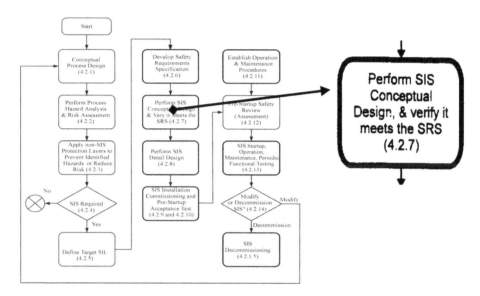

Figure 5.1
Conceptual design stage in the ISA safety life cycle

Some points taken from ISA S84.01 Clause 6 will provide a good indication of what issues should be considered at this stage.

Objectives

'To define those requirements needed to develop and verify an SIS Conceptual Design that meets the Safety Requirements Specifications.'

Conceptual design requirements

Clause 6.2.1 requires that 'The Safety Instrumented Systems (SIS) architecture for each safety function shall be selected to meet its required Safety Integrity Level (SIL). (e.g. the selected architecture may be one out of one 1oo1, 1oo2 voting, 2oo3 voting, etc.)' This is an important feature and is one that has become a significant feature of the IEC standards 61508 and 61511 where the architecture of each part or subsystem of the SIS must comply with certain minimum requirements for fault tolerance. We shall examine these in more detail in Chapter 7

Clause 6.2.2 states, 'A SIS may have a single safety function or multiple safety functions that have a common logic solver and/or input and output devices. When multiple safety functions share common components, the common components shall satisfy the highest SIL of the shared safety function. Components of the system that are not common must meet the SIL requirements for the safety function that they address.' This is a fundamental rule for all safety system designs

Clause 6.2.3 provides a very useful list of the design features that can be used to ensure that the SIS can meet the required SIL rating. The list of features is given below and readers are encouraged to make reference to the ISA standard to follow up on the guidance material provided by the ISA standard in its appendix B.

a)   Separation – identical or diverse (see B.1 for guidance)

b)   Redundancy – identical or diverse (see B.2 for guidance)

c)   Software design considerations (see B.3 for guidance)

d)   Technology selection (see B.4 for guidance)

e)   Failure rates and failure modes (see B.5 for guidance)

f)   Architecture (see B.6 for guidance)

g)   Power sources (see B.7 for guidance)

h)   Common cause failures (see B.8 for guidance)

i)   Diagnostics (see B.9 for guidance)

j)   Field devices (see B.10 for guidance)

k)   User interface (see B.11 for guidance)

l)   Security (see B.12 for guidance)

m)   Wiring practices (see B.13 for guidance)

n)   Documentation (see B.14 for guidance)

o)   Functional test interval (see B.15 for guidance)

The design features given in Appendix B to ISA S84.01 are clearly and succinctly described. It is noteworthy that most of the advice in Appendix B has been incorporated into the new IEC 61511 standard. Part 2 of IEC 61511 is entitled 'Guidelines in the application of part 1' and is due for release in 2003. We will give more practical details on field instruments and engineering later in this book but here are some key design points from the Annex B paragraphs that are relevant to the conceptual design stage.

Key points on separation of safety systems from control systems.

- Separation – identical or diverse. Applicable to BPCS and SIS. SIL 1 systems can accept identical separation, diverse preferred for SIL 2 and SIL 3
- Separation applies to field sensors, final control, logic solver, and communications. For example: control valves. SIL 1 accepts a shared valve for isolation; SIL 2 prefers a separate valve. SIL 3 calls for identical or diverse separation
- Logic solver: SIL 1 single separate. SIL2 and 3 identical or diverse separation
- Special conditions for integrated safety and control systems.

Key points on redundancy

- For avoidance or minimizing of spurious trips use redundancy
- Take care to avoid common cause faults in redundant designs
- Use the advantage of diverse redundancy in sensors

Key points on technology

- Relay systems, merits and demerits, design basics
- Electronic technology comments, e.g. timers
- Solid state logic, not recommended except with diagnostics or as pulsed logic
- PES comments as per later in this chapter

Key points on architecture

- Fail safe philosophies, diverse/redundant choices, redundant power sources, operator interface, and communications
- SIL 1 acceptable to use single channel architecture
- SIL 2 more diagnostics plus redundancy as needed
- SIL 3 diverse separation, redundancy and diagnostics are significant aspects
- User to evaluate failure rates and SIS performance to validate the design

Key points on common cause failures

Common cause faults arise when a problem is equally present in two separate systems. For example if two pressure transmitters are identical and both have been wrongly specified there is no protection by the 2nd unit against errors in the first unit. It doesn't help to have a twin-engine aircraft if you have water in all the fuel!

5.2.2   IEC 615108 on conceptual design

There is no distinct conceptual design phase in IEC 61508 but the initial design considerations mapped out in the 'safety requirements allocation' stage will require some conceptual design activity. The approach in this standard is to allocate risk reduction duties to the SIS and then develop the design properly when the project reaches the detail design phase. However it is reasonable to expect that the outline of a practical design will normally be prepared before the detailed safety requirements specification has been drawn up. Hence the design concepts with most of the key features will be drawn up as early as possible in a project to establish the feasibility of the safety function.

Detailed hardware design requirements are set out in part 2 of IEC 61508, covering the system realization stages. We are going to look at this in more detail in Chapter 7. For the purposes of overall system design at the conceptual stage the principles laid down in IEC 61508 are essentially the same as ISA S84.01 Annex B. Again it is worth noting that a valuable list of good design practices and considerations will be found in part 2 of IEC 61511 when that part of the standard is finally issued.

5.2.3   Skills and resources

As noted in the previous chapter IEC 61508 clause 7.6.2.2 calls for the skills and resources available during all phases of the safety life cycle to be considered when developing the overall safety system including the SIS. There is a comment to the effect that a simpler technology may be equally effective and have the advantages of reduced complexity. This is a sensible reminder that we should not propose a 'high tech' solution for a 'low tech' environment.

5.2.4   Conceptual design stage summary

Once the decision has been made to consider a safety instrumented system for a protection function the conceptual design stage involves the following basic steps:

- Define the safety function and required SIL

- Decide the feasibility of measuring the parameters that will signal the need for a shutdown action
- Decide the feasibility of final elements to achieve the shutdown
- Establish the process safety time and check that the sensors and final elements can operate well within that time
- Outline the architecture requirements to achieve the SIL and to provide adequate protection against spurious trips
- Decide on the type of logic solver system that is most suitable for the application bearing in mind the number of safety functions that are needed for the process plant, the complexity of the functions, the technical resources available to the plant and the cost of ownership
- Review the impact of the logic solver capabilities on the choice of sensing and final element devices and carry out a preliminary reliability analysis to confirm that the SIL target can be met. Revise the design as necessary
- Produce a summary report on the conceptual design and file this with the records of the safety allocations phase (IEC phase 5). Use this report as a reference for the safety requirements specification to be prepared in phase 9.

We should get the basic engineering right early in the project. Don't put off basic thinking or evaluation until the whole scheme has to be designed, ordered and delivered.

Let's look at some basic SIS configurations and see what options we have for the technology of the logic solver. We shall look at the sensors and actuators in Chapter 7.

5.3   Technologies for the logic solver

In this section we review the essential features of logic solver technologies in the context of conceptual design. Some of the details here may be vendor specific but this does not imply any particular preference for a given product.

We begin by revisiting the basic configuration of a safety instrumented system to help us to recognize the role of the logic solver.

5.3.1   Basic SIS configuration

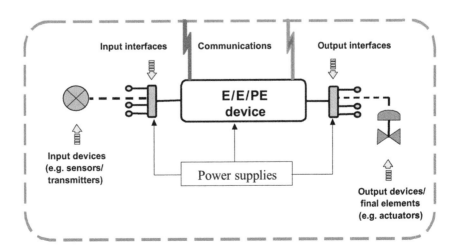

Figure 5.2
Basic SIS configuration

The basic SIS as shown in Figure 5.2 will generally be comprised of the following:

- Sensor or sensors with associated signal transmission and power
- An input signal processing stage
- A logic solver with associated power supplies and a means of communication to either an operator interface or another control system
- An output signal processing stage
- Actuators and valves or switching devices to execute the final control element function

This chapter describes the typical PES based on PLCs or specialized processor modules but even a simple relay based shutdown system has the basic parts listed above.

5.3.2   Shared functions

As soon as more than one SIS function has been identified (specified) for a process the question arises: Should each SIS be completely individual? In most cases the answer for reasons of cost and practicality is usually No. There is very often a need for multiple safety functions to share the logic solver and all its facilities such as the interface and the power supply. The standards allow this provided the safety integrity of each individual safety function is evaluated. So the architecture model for the SIS can be modified as shown in the next diagram:

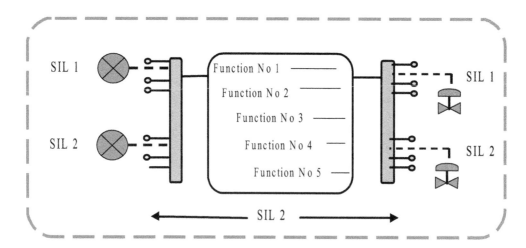

Figure 5.3
Shared functions in the logic solver: highest SIL applies

It is important to note here that the safety integrity of the logic solver (or at least the shared parts of it) must be rated to satisfy the highest SIL of the shared functions. This is a significant point to consider at the time of selecting a logic solver system. For example, it typically happens that a plant will have 95% of its safety functions rated for SIL 1 and SIL 2 and only one or two SIL 3 applications. It may be attractive to buy an SIL 2 rated logic solver for all except the SIL 3 special applications and then install a small solid-state logic unit for the SIL 3 application. Some systems offer modular hardware options to install input/output subsystems with different SIL ratings to optimize on hardware costs.

5.3.3 Technology choices

In the following paragraphs we take a look at the features of each type of logic solver technology. All the types we are considering remain valid choices because of the wide range of situations that make use of safety instrumentation.

5.3.4 Pneumatics

Pneumatic devices have been used extensively in the offshore oil industry and continue to be used in freestanding installations in petrochemicals where the great advantage is that they are inherently safe for hazardous atmospheres. There is nothing in the design guidelines for SIS that would stop a pneumatic system from being used as a low integrity SIS.

The most common application is to use a field mounted pneumatic controller to provide wellhead pressure protection. These units compare the delivery pressure with a set point. The controller output signal goes to a pressure switch that drives a final element to execute a delivery valve closure.

5.3.5 Relays

Relay based shutdown systems were the mainstay of the process industry up until the arrival of solid-state systems in the 1980s.

Figure 5.4 shows the classic features of a relay based shutdown system. The match with the generic model in Figure 5.1 is easy to see where the main features are pointed out.

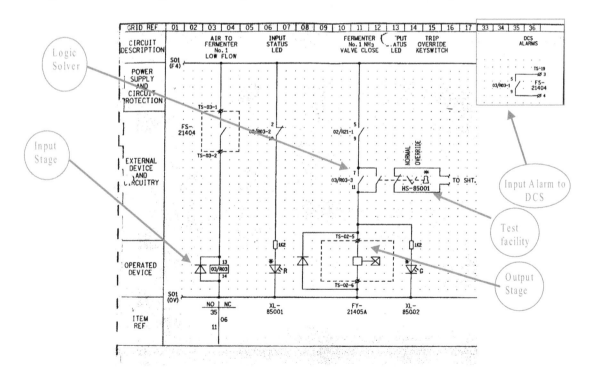

Figure 5.4
Relay based shutdown system

Relay systems are simple to design with a good safe failure proportion by using relays with normally open contacts. They are well suited to simple logic applications but have the disadvantage of requiring a comparator or trip amplifier to generate the logical state values from analog transmitters. Figure 5.5 illustrates the principle.

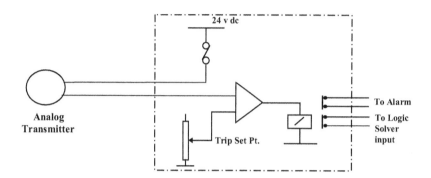

Figure 5.5
Principle of the trip amplifier as an input converter to the relay logic

Similarly the relay based logic solver is unable to carry out any form of calculation function without the use of special purpose computing modules. These have often proved difficult to set up and calibrate and are essentially obsolete in comparison with present day computing power.

Relay systems will always have a place in safety systems and should be carefully considered for simple applications. The following table lists the merits and de-merits may be considered in making a choice for or against relay based systems.

Good Points	**Bad Points**
Simple and Inexpensive for Small Systems.	Complexity grows quickly.
EMI and RFI immune.	Needs trip amplifiers for analog signals: Prone to failures and RFI.
Low tech. Useful for basic plants. Hardwired.	Difficult to modify reliably: Wiring details obscure and prone to errors.
Generally fail safe: Failure modes and failure rates can be quantified.	Prone to nuisance trips. Trip amps are a weakness.
Easy to service.	Too easy to corrupt. Poor security.
Adaptable.	Custom built: Quality assurance problems, prone to design errors Expensive to build larger systems.
Easy to design when simple.	Expensive to document.
Easy to manually test.	No diagnostics.
	Poor communications/interfacing.

Table 5.1
Merits and de-merits of a relay based logic solver

5.3.6 The safety relay

Whilst we are considering relay based systems its is important to be aware of the wide range of applications and devices in relays applied to machinery safety systems. In simple applications such as emergency stops and guard interlocks the safety integrity of the protection system depends very often on a single relay or a pair of relays and a pair of field contacts.

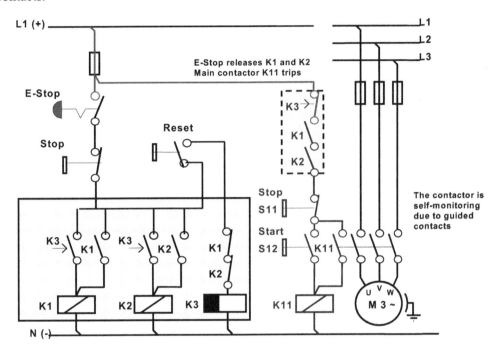

Figure 5.6
Typical safety relay arrangement for an emergency stop function

The requirements for a high integrity switch input and relay logic device led to the development of the safety relay or monitoring relay module. This is a modular assembly of relays arranged to operate as a dual redundant pair with a third relay as a self-checking or diagnostic function. The arrangement provides a high degree of assurance that the switching function will be available due to the redundant pair of relays as well as the self-testing that takes place each time the unit is energized.

Figure 5.6 shows a typical application to an emergency stop function. The monitoring relay assembly will not reset if any of the relays is not in its correct state at the time of start up. The integrity of the safety relay depends on the principle of 'positively guided' contacts. This requires that the contact sets in the relay be directly and rigidly linked to each other. Then it becomes almost certain that the state of one pair of contacts will always define the state of the other contacts.

Safety relay modules can be of value in process safety systems because of their high integrity and redundant characteristics. They may be used as input stage devices but will be expensive to use for the logic functions. The self-test on start up is of limited value in low demand applications where reset may only take place once a year.

5.3.7 Solid-state systems

At the same time as the need for more complex and reliable shutdown systems became pressing so the availability of smaller and smarter electronics increased. The era of solid-state systems as an alternative to relay based systems probably began in the late 1970s and ran until the establishment of really attractive programmable system solutions in the mid 1990s. From the evidence of the applications still being installed it looks as if the best of the solid-state systems is going to continue to be used and improved for many years to come.

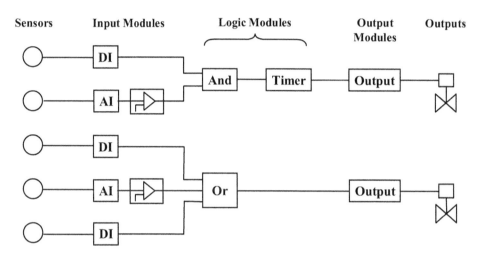

Figure 5.7
Elements of a solid-state logic solver

This diagram shows the configuration of a typical solid-state SIS with its input signal processing stage; the logic solver function being performed by standardized electronic function blocks mainly AND gates, OR gates, logic inverters and timers.

Considering the merits and de-merits of solid-state systems they have essentially the same characteristics as relay based systems with the advantage of using purpose built components such as multi-channel input signal processing boards and logic solver blocks. Early versions suffered from a substantial disadvantage over relays because they lacked fail-safe capabilities. It is possible for a static logic element to fail high or low or just stop switching. The failure may remain undetected and hence presents a high fail to danger risk.

The answer to the failure mode question was to use dynamic logic. The modules of the logic solver are operated in a continuous switching mode transmitting a square wave signal through each gate or circuit. Diagnostic circuits on board each module then immediately detect if the unit stops passing the pulses. The detectors in turn link to a common diagnostic communication module that reports the defect to the maintenance interface. Normally the detection of a failed unit will lead to an alarm and sometimes a trip of the plant.

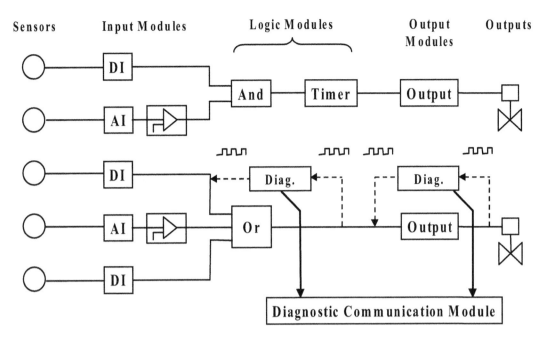

Figure 5.8
Solid-state systems: diagnostic communication module

Figure 5.8 illustrates the principle of injecting a cyclic switching signal to each stage of the logic to provide a continuous diagnostic on the correct functioning of each stage.

Here is a brief run through of the main features of a currently available solid-state system made by the one of the leading specialist companies in the field: HIMA.

Features

- Safety related hardwired controls up to Requirement Class 7/SIL4.
- Self-diagnosis function on each module.
- Easy localization and replacement of failed modules.
- Diagnostic and communication module.
- Communication with DCS or PES via the communication module (for MODBUS, Ethernet [10BaseT], RS 485).

Intelligent technology

- Modules in 19' European format with 3 units high and 4 units space to DIN 41494.
- Plug connectors to DIN 41612, version F.
- Operating voltage 24 V = / –15 % ... +20 %.
- Temperature range –25° C ... +70° C.
- Operating voltage L – earthed or unearthed.
- Use of standard wiring techniques.

Diagnostics

- Self-diagnostics in all modules with the diagnostic/communication module (DCM)
- Signaling of faulty modules by the red LED on the front of the module

- Assembling the fault signals by one common signal or in groups by means of the signal contacts of the modules
- Line monitoring of input modules with signaling and LED
- Fuse monitoring of output modules with signaling and LED

Communication

- Information about type of module, input and output signals as well as fault reporting.
- Event recording (change of binary signals with time).
- One communication module collects the information of one sub rack with 20 modules.
- MODBUS communication without any additional configuration software.

Solid-state system components

Module Types

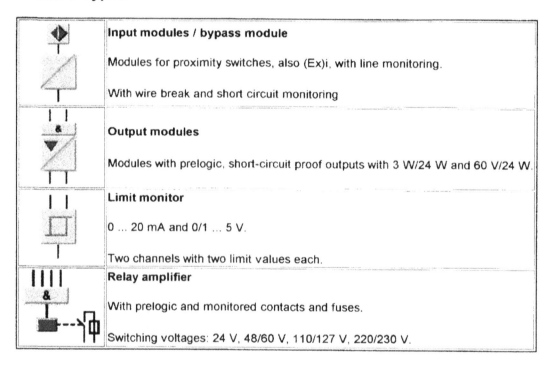

	Input modules / bypass module
	Modules for proximity switches, also (Ex)i, with line monitoring.
	With wire break and short circuit monitoring
	Output modules
	Modules with prelogic, short-circuit proof outputs with 3 W/24 W and 60 V/24 W.
	Limit monitor
	0 ... 20 mA and 0/1 ... 5 V.
	Two channels with two limit values each.
	Relay amplifier
	With prelogic and monitored contacts and fuses.
	Switching voltages: 24 V, 48/60 V, 110/127 V, 220/230 V.

Figure 5.9
Typical solid-state system components (from HIMA publications)

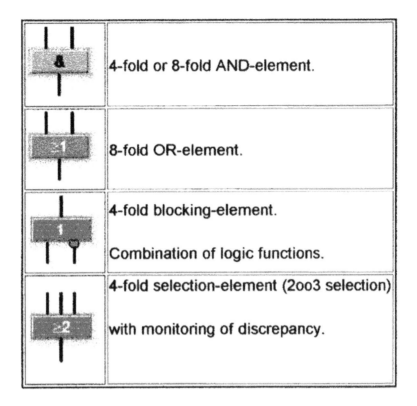

	4-fold or 8-fold AND-element.
	8-fold OR-element.
	4-fold blocking-element. Combination of logic functions.
	4-fold selection-element (2oo3 selection) with monitoring of discrepancy.

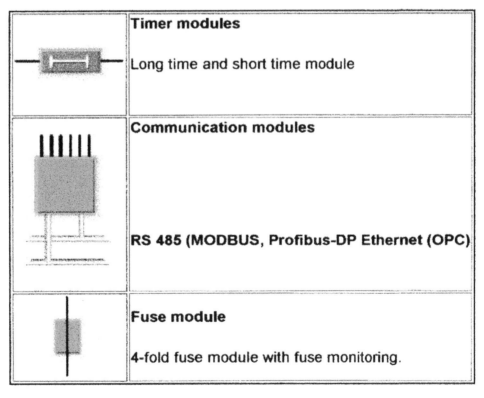

	Timer modules Long time and short time module
	Communication modules **RS 485 (MODBUS, Profibus-DP Ethernet (OPC)**
	Fuse module 4-fold fuse module with fuse monitoring.

Figures 5.10 – 5.11
Typical solid-state system components (from HIMA publications)

Component features

- Safety related modules, tested on DIN V 19250 and IEC 61508.
- Certified for use up to AK7 (DIN V 19250) and SIL4 (IEC 61508).
- Modules with microprocessors (limit monitor, time level) are approved up to AK6/SIL3.
- CE certified.

1st level diagnostics

- Diagnostics and communication module on each module
- Provides common information, status IO signals, current values and presets

2nd level diagnostics

- Communication module in each rack
- Polling of the data of all the modules in the rack
- Generation of events (change of I/O signals)
- Data pool for external communication

3rd level diagnostics

- Master system (MODBUS, PROFIBUS-DP) or OPC server to request common information, status IO signals, current values, presets and events

Fields of application

- For high safety requirements.
- For very high availability requirements.
- For high degree of module stability.
- High Integrity Pressure Protection Systems (HIPPS),
- Emergency shutdown systems, burner control systems and fire & gas systems.

We should note here that one of the areas where these systems seem to be popular is where a machine such as a large gas compressor or turbine is used in repeated copies and where a very high level of safety integrity is needed. It follows that the engineering costs of the application can be spread over a number of plants and the protection logic is well defined and stable. Thus for example in gas field distribution projects the main process safety system is likely to use a PES whilst the compressor stations may be built with fixed function solid-state systems.

The following table summarizes the merits and demerits of the solid-state logic solver option.

Good Points	Bad Points
Simple and Inexpensive for Small Systems.	Complexity grows quickly.
EMI and RFI immune.	Needs trip amplifiers for analog signals: Prone to failures and RFI.
Low tech. Useful for basic plants. Hardwired.	Difficult to modify reliably: Wiring details obscure and prone to errors.
Generally fail-safe: Failure modes and failure rates can be quantified.	Prone to nuisance trips. Trip amps are a weakness.
Easy to service.	Too easy to corrupt. Poor security.
Adaptable.	Custom built: Quality assurance problems, prone to design errors. Expensive to build larger systems.
Easy to design when simple.	Expensive to document.
Easy to manually test.	No diagnostics.
	Poor communications/interfacing.

Table 5.2
Characteristics

5.3.8 Programmable systems for the logic solver

Programmable systems, in particular the safety PLC have become the most widely used devices for the logic solver duty in the process industries. Before we go any further into the design of safety PLCs let's consider what benefits we are looking for.

What are the potential benefits of using a PLC for safety?

If we wanted to justify the PLC against any other method of implementing safety, what can it offer? Some of these are likely to be the same as the benefits of a standard PLC over hardwired systems, such as:

- Software tools for configuration and management of the logic.
- Simplified wiring eliminates the problem of logic being embedded in the connections between hardware modules.
- Using software for safety functions allows the building of machines to be completed whilst the final protection logic is being developed. Facilitates late design developments and avoids wiring modifications.
- Standard control packages become cost effective when machines are produced in quantities. Customized variations can be implemented with minimal costs.
- Centralized monitoring and display facilities.
- Improved co-ordination for large production lines or complex machine functions.
- Event recording and retrieval.

In particular the safety PLC offers improved safety performance through:

- Improved management of safety functions. This is due to the strict control of application software through the programming tools. Unauthorized access to the software is prevented by password control. All changes required a double compilation task. All changes are recorded.
- Pre-approved software function blocks provide standardized methods for routine safety functions.
- Powerful diagnostics for the detection of faults in the PLC and its I/O subsystems.
- Application blocks include for diagnostics to be performed on the sensors.
- Easier certification through the use of safety certified PLCs with certified function blocks.

Productivity improvements will be sought for large installations through:

- More advanced logic and sequencing capabilities to reduce lost time in safety functions.
- Better testing facilities.
- Rapid detection and location of faults.

Bus technologies interfacing into some PLCs further improve cost effectiveness through:

- Allowing plant wide safety functions to be managed from central or remote stations.
- Remote I/O sub systems reduce cabling and reduce risk of cabling errors.
- Safety certified field bus systems allow a single bus connection for all safety sensors on a machine, reducing wiring complexities.

What are the potential disadvantages of using a safety PLC?

Probably the main disadvantages are associated with capital cost when considering a PLC for a relatively simple application. Safety PLCs are much more expensive than standard PLCs and the cost of the software package and any special training must be added in. If the application is to be repeated many times the cost equation will become more favorable as the software investment is recovered. For a simple application the modular products we have seen will be a cheaper solution until the safety function becomes complex. For example there are a number of packaged PLC solutions on the market for the complete safety functions for mechanical and hydraulic power presses. The scope of these solutions as a contribution to increased safety as well as higher productivity may well justify their capital cost.

Further reservations as far as the end user is concerned may arise from the need for more specialized knowledge on the part of the maintenance team. Again the benefits of improved diagnostics and testing facilities may offset this concern.

5.4 Development of safety PLCs

5.4.1 Why not use general purpose PLCs for safety functions?

Standard PLCs initially appear to be attractive for safety system duties for many reasons such as those listed here:

Attractions

- Low cost
- Scalable product ranges

- Familiarity with products
- Ease of use
- Flexibility through programmable logic
- Good programming tools available
- Good communications

The PLC fits in easily to the SIS model as shown here.

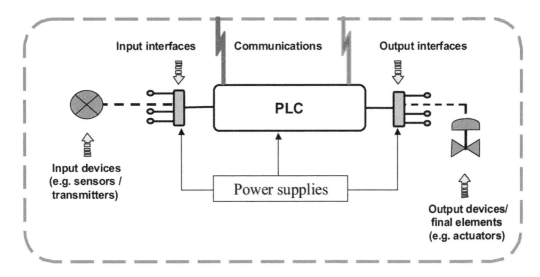

Figure 5.12
General arrangement of a programmable system in a safety instrumented system

But there are significant problems.

- Not designed for safety applications
- Limited fail safe characteristics
- High risk of covert failures (undetected dangerous failure modes) through lack of diagnostics
- Reliability of software (also stability of versions)
- Flexibility without security
- Unprotected communications
- Limited redundancy

For process safety, SIL 2 and SIL 3 normally demand a fault tolerance level of at least 1 or at least very high diagnostic coverage. (See Appendix A). The risk of hidden dangerous faults and the complexities of arranging dual redundant architectures increase the engineering cost and complexities for those wishing to adapt standard PLCs to meet these requirements.

So if we want to use a general purpose PLC it has to be specially engineered to meet the basic requirements for safety duties. Consider for example the need for I/O stage diagnostics:

I/O stage diagnostics

Here's a simple example of the covert failure problem. The output stage of the PLC operates a fail-safe solenoid or motor trip relay. It may have to stay energized for weeks but we won't know if it is shorted until it has to trip the function. This is an unrevealed fail to danger condition or 'covert fault'. The broken wire fault is an 'overt fault' or revealed fault, which will fail to a safe (off) state but creates a 'nuisance trip'.

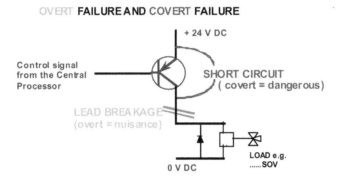

Figure 5.13
Failure modes of a PLC output stage

Some users of standard PLCs have been able to introduce their own diagnostics for continuously self-checking the PLC whilst it is on-line. For example see Figure 5.14:

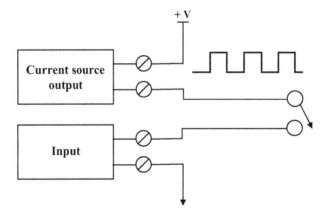

Figure 5.14
Simple diagnostic for PLC input stage

For output switching stages a typical method of self-testing consists of a pulsed off state that is too short to affect the load but which can be read back into an input stage as part of a test cycle. See figure 5.15 below.

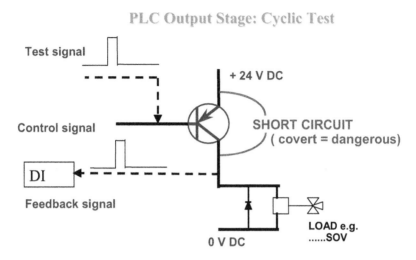

Figure 5.15
PLC output stage: cyclic test

Another approach for assuring input stage integrity is to use voting as a method of diagnosing a fault. In the next figure the digital input is not accepted as valid unless a majority of 2 out of 3 input channels agrees on the state (this a 2oo3 architecture). If one channel disagrees with the other two the whole input stage can be treated as faulty or the fault can be reported and the PLC continues to operate on the majority vote until a repair is made.

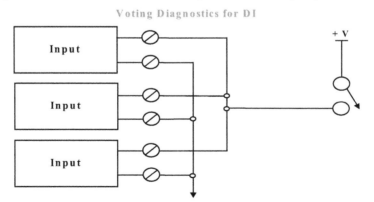

Figure 5.16
2oo3 voting diagnostics for a digital input function

Overall PLC reliability

What we have seen so far are some simple measures to improve the safety integrity of the standard PLC. The problem for us is that the list of potential failure modes is a long one and the cost of avoiding or safely detecting each one is rising.

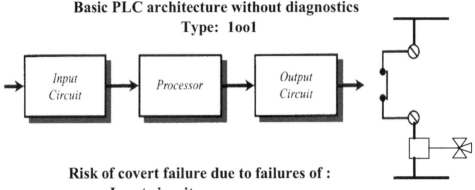

Basic PLC architecture without diagnostics
Type: 1oo1

Risk of covert failure due to failures of :
* **Input circuits**
* **I/O comms**
* **Processor**
* **Program cycle**
* **Output circuits**

Figure 5.17
Basic PLC architecture without diagnostics

According to ISA S84.01 there are at least 36 recognized types of failure modes in a typical PLC and another 6 failure modes in the I/O sub systems. Whilst, some of these failures are rare, they are not acceptable for safety systems. What we are looking for is a situation where at least 99% of all possible failures can be certain to result in a safe condition for the machines and persons being protected by the PLCs. So on the grounds of hardware performance alone engineers are faced with a tough task to make the standard PLC do the job.

Software reliability considerations

There are similar reservations about the suitability of standard PLC software for safety:
* Potential for systematic faults
* Program flow and monitoring is not assured
* Too much scope for random applications (high variability)
* Lack of security against program changes
* Uncertain response in the presence of hardware faults
* We can't test for all foreseeable combinations of logic
* The failure modes are unpredictable.
* Re-use of old software in new applications (absence of quality trail)

Fundamentally the problem with using software in a safety system lies with the potential for systematic faults: i.e. faults that are not random such as component burnouts but have been introduced during the specification, design or development phases by errors in those processes. Such faults may then lie dormant until just the right combination of circumstances comes along.

One of the aspects of software that makes it particularly difficult to control is the ability for it to be re-used in new applications not originally intended by the designers. The tendency to use 'cut and paste' techniques to make up a new program creates a risk that the wrong features have been introduced to a new product.

It would help if it were possible to detect all systematic errors at the testing stages but it is well understood that there will be many combinations of logic and timing that cannot be fully explored in testing without a prohibitive cost in time and labour.

Do all these objections eliminate standard PLCs?

No, they are not eliminated, but they are likely to be the most expensive route to go. There are many existing applications where standard PLCs have been adapted to safety related control functions. These applications have involved installing dual redundant sets of PLCs and a number of self-checking measures along the lines we have shown. In some cases the recognized certification bodies for a specific application have approved these solutions.

However with the availability of specially engineered safety PLCs for general use in industry it becomes far less attractive on grounds of cost and engineering effort to take that route. Even when the system has been adapted for safety it still requires a third party to verify that it is suitable for the safety duties.

5.4.2   Upgrading of PLCs for safety applications

Summarizing the position: It is possible to upgrade software engineering through improved QA techniques. It is possible to consider dual redundant standard PLCs in hot standby mode but the standard PLC does not lend itself to covering all the possible failure modes through the normal fault detection systems. We can add our own diagnostic devices for some types of failures as we have seen. But at the end of all these extra efforts we have the problem that we have built a special application that needs to be carefully documented and maintained. And then we have the problem of proving it to others or certifying it for safety duties.

In a review of the position regarding standard PLCs compared with safety PLCs, industry specialist Dr William M. Goble concluded:

'The realization of many users that conventional controllers cannot be depended upon in critical protection applications creates the need for safety PLCs. The standards are high for safety PLC design, manufacture and installation. Anything less than these high standards will soon be considered irresponsible, if not negligent, from a business, professional and social point of view.' From a paper by Dr. William M. Goble, Exida, www.exida.com

This is basically the case for vendors to produce a special purpose PLC built specifically for critical safety applications. Lets look at what it takes.

5.4.3   Characteristics of safety PLCs

In summary:
The answer to the problem of undetected faults in PLCs lies in the concept of Fault Coverage and Fault Tolerant Systems: The answer to the problem of hidden defects in software is high quality embedded (i.e. operating system) software combined with strictly defined and constrained user programming facilities.

5.4.4 Hardware characteristics of a safety PLC

- Automatic diagnostics continuously check the PLC system functions at short intervals within the fault tolerant time of the process.
- High diagnostic coverage means that at least 99% of all hardware faults will be detected and notified for attention and repair.
- Provides a predictable and safe response to all failures of hardware, power supplies and system software.
- Redundant hardware options available to provide safe operation even if one channel has failed.
- Fault injection testing of the complete design is performed to ensure safe failure response to all known faults.
- I/O sub systems continuously check all signal channels. I/O bus communications are self-checking; faults result in safe isolation of affected I/O groups.
- High security on any reading and writing via a digital communications port.

5.4.5 Software characteristics of a safety PLC

Software quality assurance methods are deployed throughout the development and testing of both operating system and application software development. Software development takes place under 'safety life cycle' procedures as specified in IEC 61508 part 3. The software design and testing is fully documented so that third-party inspectors can understand PLC operation.

Operating system uses a number of special techniques to ensure software reliability. These include:

- 'Program flow control' checking, this insures that essential functions execute in the correct sequence.
- 'Data verification' stores all critical data redundantly in memory and checks validity before use.
- Operating system and user application software tools are approved for safety by third party approval bodies.
- Operating system and programming package supplied by same vendor as the hardware.
- Software and hardware integration tested by approval bodies.
- Extensive analysis and testing carefully examines operating systems for task interaction.
- Application software uses 'limited variability languages' to restrict end users to working within a framework of well-proven instructions and function blocks.
- All application software updated transparently to redundant channels.

Whilst all of the above are general performance and qualification features of the safety PLCs, the practical end user will also be interested in some more down to earth characteristics. For example users will look for:

- Economically priced PLCs at the right size for typical process applications.
- Input channels suitable for all common safety sensors and output channels suitable for connection to secondary or final control contactors or solenoids.
- Remote I/O capabilities to allow input and output modules to be mounted close to the parts of the machine or production line that they serve.

- Speed of response fast enough to deliver E-Stop and safety trip responses without increasing risks to persons.
- Low software engineering costs, library of certified safety function blocks.
- Easy to program with fill-in-the blanks function blocks plus simple ladder logic or sequential logic instructions. Program language should be as close as possible to the type in use for basic control PLCs.
- Good testing and diagnostic facilities.
- Rapid identification of faulty parts and easy replacement.
- Compatible but safe connections to automation control networks.

Generally it will be best if the safety PLC is, in all respects except safety, the same product for the end user as the standard PLC.

5.4.6   Design of safety PLCs

This section illustrates some of the features typically found in safety PLCs. We begin with a single channel system and work towards more sophisticated designs. We can only provide here a brief look at the key features and would advise that it is generally possible to track these features in greater detail by close study of the technical descriptions of the products provided by the various manufacturers.

We shall see in this section that different types of safety PLCs are evolving to suit the type of industry applications. The major division is between process industry applications and machinery safety.

Single-channel safety PLC architecture with diagnostics

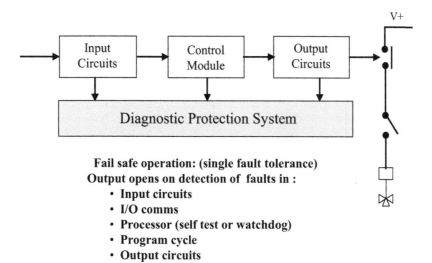

Figure 5.18
Single channel PLC architecture with diagnostics type – 1001D

The above diagram shows the basic architecture of the PLC upgraded to include for diagnostic devices embedded in the construction of the PLC. This unit is able to overcome the objections listed for standard PLCs and is now the basic module concept for several safety control system manufacturers. Essentially this unit in single channel configuration will trip the machinery if any of its diagnostic functions finds a fault in any of the stages: Input, CPU, power supply or output.

The term 1oo1 comes from the notation that any 1 dangerous fault in the 1 channel system will cause a failure of the safety function. The D denotes diagnostics protection. (It may be helpful to refer to appendix A.).

There are two important features to note here:

- Diagnostics in the single channel system must be performed within the PLC cycle time such that the 'fault tolerance time' or 'process safety time' is not exceeded. (See Figure 5.19.)
- The diagnostic circuits must have a means of shutting down the outputs of the PLC that is independent of the output switching circuits. This is sometimes known as a 'secondary means of de-energizing' (SMOD).

Process safety time

The process safety time is defined as the period of time between a failure occurring in the EUC or the EUC control system (with the potential to give rise to a hazardous event) and the occurrence of the hazardous event if the safety function is not performed. It follows that a safety system must perform its measurement and logical responses in less than the process safety time. Some of the spare time available within the process safety time can then be utilized to perform diagnostic checks as illustrated in Figure 5.19.

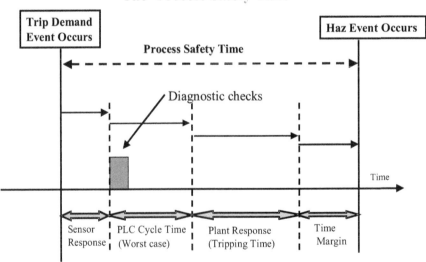

Figure 5.19
Fault tolerant time or process safety time

All self-diagnostics are executed typically within a 1-second time frame. TUV class 3 specifications allow a 1-second interval to execute the following:

- Diagnostics to ensure very high coverage of possible faults
- Application software executed at least twice
- Operate a secondary means of de-energization of all 'fail-safe' outputs in the event of diagnostics finding a fault

Diagnostic coverage

An essential requirement of the diagnostic tests for the PLC is that they should cover a high percentage of potential faults. Diagnostics linked to appropriate control actions convert potentially dangerous hidden faults into safe mode failures of the PLC. If a substantial number of potential faults remain that cannot be detected by diagnostics then much of the benefit is lost.

'Diagnostic coverage is the ratio of safely controlled faults to all possible faults'

from Steven E Smith: Fault Coverage in Plant protection Systems, ISA 1990 Paper #90-363....

This paper provides an excellent review of fault coverage techniques.

1001D safety PLC

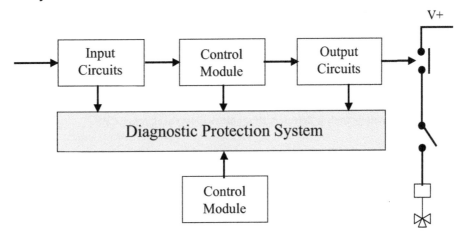

Figure 5.20
Single channel safety PLC architecture with dual CPU

Here is an upgraded version of the single channel module where availability is improved by adding a second hot standby CPU to allow continued operation if one of them fails. This is still a single channel system.

The problem with a single channel safety PLC is that since by definition no single fault must leave the SIS function unprotected it must always cause a trip when a fault is detected. Consequently the availability may suffer; i.e. the nuisance trip rate increases. The answer to this is to go to the Dual 1002D configuration shown in the next diagram.

1002D safety PLC

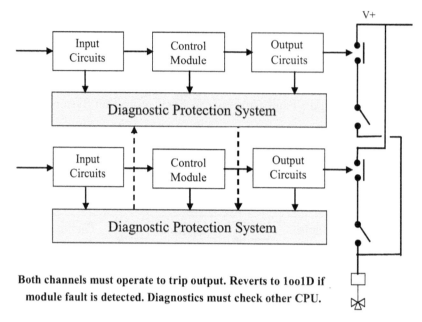

Both channels must operate to trip output. Reverts to 1oo1D if module fault is detected. Diagnostics must check other CPU.

Figure 5.21
Dual redundant channel safety PLC architecture with diagnostics

Parallel-connected type: 1oo2D

In this version the entire logic solver stage from input to output is duplicated and if one unit fails its diagnostic contact will open the output channel and remove that unit from service. The SIS function then continues to be performed by the remaining channel.

The notation 1oo2D applies because the system will still perform in the presence of 1 fault amongst 2 units. The parallel connection of the two units substantially improves the availability. Note that diagnostic performance is further improved by cross-linking between the CPU of one channel and the diagnostics of the second channel.

Series-connected type: 1oo2

Series connection of the outputs of two PLCs means that either of the channels can trip the plant without depending on the diagnostic sections. This configuration is not popular except for unusually high safety integrity requirements. It does however raise the point that through modular construction the vendors of these systems can offer numerous combinations of modules with selective redundancy to suit specific applications. For example a burner management package may have 1oo2D configuration for critical safety functions and 1oo1 for other less critical functions all built into the same assembly of racks.

Common cause potentials in redundant PLCs

Despite the practice of high quality hardware and software engineering there is always a small possibility that two identical PLCs connected in redundant mode will both suffer the same common defect. Whilst co-incidence in time seems highly improbable for hardware defects the possibility of a systematic fault in the embedded software cannot be ruled out. The next two diagrams give an example of a technique deployed by Siemens-Moore designed to virtually eliminate the risk of common cause defects in the software of their Quadlog system.

Mode switching PLCs

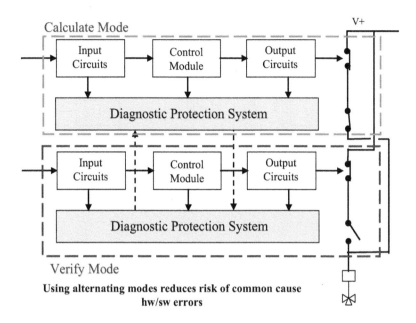

Figure 5.22
1oo2D mode switching (calculate/verify) PLC pair

In this application one of the 1oo2D pair of PLCs switches its output off whilst the other performs normal duties in what is called 'calculate mode'. For a period of typically 12 hours the arrangement stands and the 1st unit operates in 'verify' mode whereby it operates a different program designed to monitor and verify the operations of the 'calculate' PLC. At the same time it still has its own diagnostic processor checking itself and its partner.

After the given period the units exchange modes and the plant protection duties are transferred to the output controls of the 2nd unit. The result of this alternating duty cycle is that two different software programs are protecting the plant in alternate processors. The potential for common mode failures is reduced by at least an order of magnitude.

5.4.7 Triple modular redundant or TMR systems

The safety systems built on the 1oo2D modules have found a strong market in the general area of process plant applications. In some of the most demanding safety areas including offshore oil and gas and in the nuclear field these systems have to compete with the alternative architecture based on the principle of 2 out of 3 voting. These are known as triple modular redundant or TMR systems. The next diagram illustrates the principle.

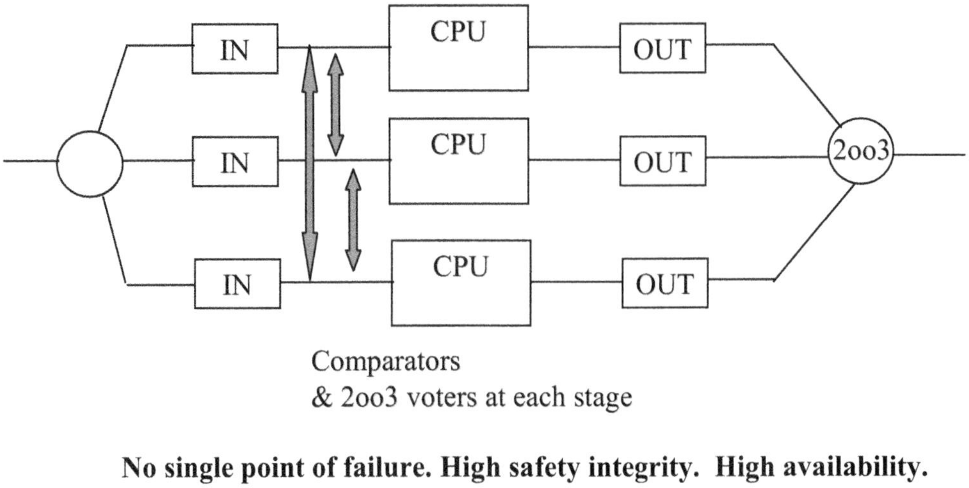

Comparators
& 2oo3 voters at each stage

No single point of failure. High safety integrity. High availability.

Figure 5.23
Triple modular redundant (TMR) safety controller system

These units have the advantage of not having to use such complicated diagnostics as the 1oo2D systems because every stage of the PES can be built up with 2 out of 3 majority voting for each function. For example the 3 input stages operate in parallel to decide which is the correct value to pass to the processor. The three processors compare data and decide which is the correct action to pass on to the output stages etc. All internal communications can use a triplicate bus. A good example of this type is the Triconex range of controller products. Let's look at the key features claimed for the TMR systems.

- No single point of failure
- Very high safety integrity
- High availability
- Transparent triplication
- Delayed maintenance capability (for remote or unmanned locations)
- Hot spare capability for I/O modules
- Sequence of event recording

The complexity and special engineering features of the TMR systems are increasingly seen to be a handicap and a price barrier for their future use. However, at present they continue to find applications in high integrity situations, particularly in oil and gas industries.

5.4.8 Safety PLC with 1oo3 architecture

We should also note the safety PLC produced by PILZ Automation that has been designed primarily for high demand mode applications in the machinery sector. This unit is also used for process automation applications but it has some distinct difference from the previous types we have seen. This system operates on 1oo3 architecture which means that three redundant processors all have to agree for the outputs to remain enabled permitting the process to continue operating.

It seeks to avoid common mode problems by using 3 different models of CPUs made by 3 different manufacturers.

It also has a different objective from the process industry TMR designs because it concentrates on making a fail-safe response by only allowing the safe outputs to remain on if all three CPUs are healthy and only if they agree the required output states of the system.

It does not provide for the machine or process to carry on running when a processor has failed.

The PILZ PSS system can be regarded as a 1oo3 architecture system with non-identical redundant CPUs. There may be some confusion here over the notation because the PILZ handbooks describe the unit as having a '3 out of 3 (3oo3) voting system'. This terminology arises from the German notation that calls for 'number of channels that must agree for the output to stay on'. In this case 3 channels must all agree. The IEC notation we have been using calls for the 'number of channels that must call for a trip for the output to trip'. This results in the term 1oo3 for this PLC.

Key features of the PILZ PSS

The basis of the PSS controllers is the three channel voting structure we have described above. The Figure 5.24 shows the arrangement.

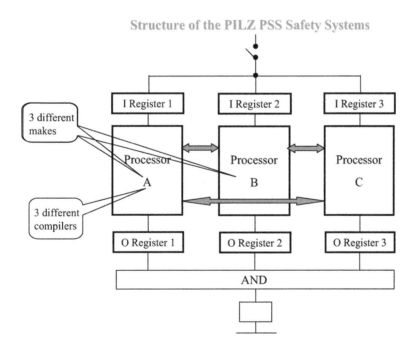

Fig 5.24
Structure of the PILZ PSS Controller

The system comprises three separate controllers, each one different from the other and supplied from different makers. Each controller has a different operating system and has its own code compiler written by different companies. These measures reduce (practically eliminate) the chances that a systematic error or hardware fault will occur as a common fault in all three controllers. The PILZ description then continues:

'Data is processed in parallel via the three controllers. Comparison checks are performed between each controller. Each processor has its own input and output register. The output register in each device is compared in an AND gate. An output will only be enabled when all three agree. This is the case for all 'bit' and 'word' functions. In other words, the PSS acts as a 3 out of 3 (3oo3) voting system.'

The PSS system has been designed to allow standard (or basic) control functions to be performed as well as safety related control functions. Safety functions are known as FS and standard are called ST. FS software runs in all three processors using an FS working bus. A separate bus is provided for the A processor to perform ST instructions. The dual function structure of the PSS has fail-safe controls operating with an independent triple voting bus whilst standard or non-safety functions operate only in processor A on a separate working bus.

Both the 2oo3 and 1oo3 types of safety PLC are particularly suited to high-speed protection systems because the voting configurations reduces the dependency on diagnostics and allow for faster cycle times. In general operators of continuous processes prefer to see 2oo3 or 1oo2D architectures due to their resistance to spurious trips. In machinery and in some batch process applications this feature is not so important and 1oo3 architectures may be considered for reasons of performance or cost.

5.4.9   Communication features of safety controllers

Before we move away from the high tech end of the safety systems we need to touch on the issue of communications. Some of the key points are summarized here:

- Strong need for communications between the SIS and the plant control systems
- For operator information and co-ordination with control
- For tidy up of DCS or PLC controller states or sequences arising from action of the SIS
- For event recording
- For I/O status and status of the SIS itself
- Security is required in communications to prevent incorrect writing of data into the SIS
- Communications and data formats need to be compatible with DCS/PLC vendor standards or open standards
- Growth of certified interfaces.

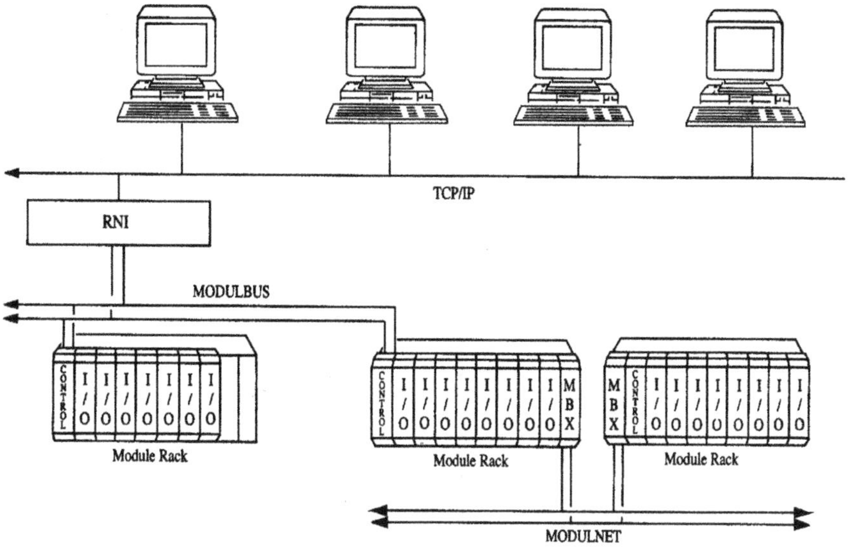

Communications example: Quadlog system with TCP/IP network for PC interfaces

Figure 5.25
Example of safety controller with network to operator interface and engineering station

Examples of communication features taken from the Siemens-Moore Quadlog System are shown in Figures 5.23 and 5.24. The description given by Siemens-Moore for the networking facilities is given below:

'COMMUNICATION WITH OTHER DEVICES AND APPLICATIONS'

'The QUADLOG critical system incorporates an open architecture that allows many other devices and applications to become an integral part of the system.'

'Existing Process Control Systems: QUADLOG has been designed to work closely with, but independently of, popular PLCs and DCSs. QUADLOG communicates seamlessly with Moore Products Co.'s APACS process control system and Honeywell's TDC 3000 (MVIP certified). In both of these situations, QUADLOG offers communication protection through the ability to identify variables not to be overwritten by any type of external communications.'

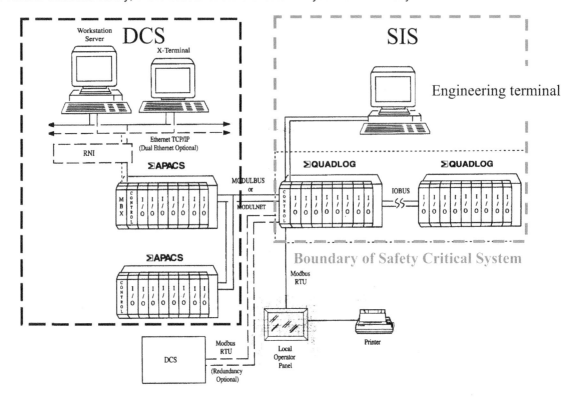

Figure 5.26
Example of DCS and SIS with integrated operator interface

The arrangement of the DCS and SIS shown in Figure 5.26 ensures that the two systems operate independently whilst being able to share information over a secure communications link. Both systems present their information to the operator on the same standard DCS workstations and displays can be integrated to serve particular process areas or functions. Access to the safety system configuration software is only possible through the dedicated engineering workstation.

Features such as these are typical for all safety controller products including TMR and solid-state systems but the critical feature is to ensure that integration of the linked systems is fully proven and trouble free.

5.4.10   New developments in communications

We can see from the above examples that a distributed architecture is acceptable for a safety PLC system provided the components are all rated for the safety duty and provided the communications network is dedicated to the safety task. It's in order to have remote I/O units with their own diagnostics linked by a safety certified network to the CPU; the network having its own diagnostics and fail to safety response.

It is also acceptable for the safety PLC to have a network interface into a DCS, a SCADA or a non-safety PLC provided this network is separately ported into the safety PLC and provided there is write protection for the safety PLC.

However, the newest development in communications goes one step further. The complete integration of safety and non-safety devices on a single network is now available in certified form using the Profibus and Profisafe protocols operating on the same network. Figure 5.27

shows the Siemens S7-400H system, which incorporates a safety certified PLC combined with a basic PLC section.

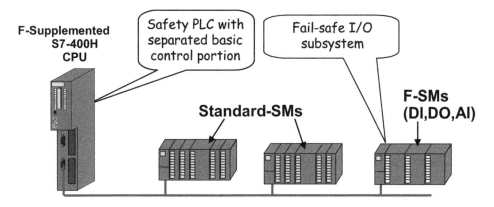

Figure 5.27
Integrated basic and safety controllers with shared Profibus/Profisafe field bus

The field connections to the safety sensors and outputs are established through dedicated safety certified input/output sub systems that carry the essential diagnostics and an independent means of shutdown in accordance with the principles we have seen. Profisafe is a certified protocol with diagnostics and fail-safe responses. The system is designed to maintain functional independence for all safety functions in hardware and software. The standard Profisafe DP protocol can then share the same network to communicate with standard controllers and standard peripheral devices such as I/O sub systems or variable speed drives.

For high availability applications the complete system can be installed with dual redundant processors, bus systems and I/O modules. This type of architecture is well suited to automation applications where many peripheral devices are interconnected.

Similarly there are established products in the automation sector using safety certified bus communications for safety devices such as emergency stops and light curtains with safety PLCs being added where needed.

These developments are not yet having any significant impact on the process sector but it is important to be aware of the potential for integrating the hardware of basic and safety controls whilst retaining all the essentials of functional independence. It is significant that the IEC 61508 standard supports this type of development by keeping all design options open to the manufacturers provided the rules of diagnostics and predictable failure modes are strictly observed.

5.5   Classification and certification

Having looked at some safety PLCs it should be clear that the technology is well established to meet the objectives of achieving high safety integrity using programmable electronic systems. The remaining obstacle is the task of proving the performance of the product. This is where the earlier DIN 19250 standard and the new IEC 61508 standard make a major contribution by setting out engineering procedures and technical performance standards that must be met in order to achieve designated SIL performance levels.

We have noted that certification must cover the hardware and diagnostic performance capabilities as specified by IEC 61508. Certification must also cover the software engineering life cycle activities leading to the embedded software as well as the programming tools supplied for the end user.

The combination of hardware and software certification means that the end user can obtain a fully certified product for the logic solver section of the SIS. This does not of course relieve the end user of any obligations as far as the quality of his software application/configuration is concerned.

Certification is a comprehensive and specialized task undertaken only by well-established authorities such as TUV. Details of testing guidelines and practices are outside the scope of this book but it is helpful to note that TUV publish details of their testing programs on their website. In particular their website publishes a list of type approved systems so that the progress of certification for each manufacturer's product can easily be checked.

5.6 Summary

This chapter has provided an outline of the issues to be faced by the designer at the conceptual design stage. Some of the basic considerations of architecture and diversity in the sensor systems will be discussed again later but the need to lay down a basic schematic for each safety function is clear. Valuable guidelines are available in the ISA standard and in IEC standards 61508 part 2 and IEC 61511 parts 1 and 2.

Technology options for the logic solver have been examined and it has been established that the safety PLC has succeeded in overcoming many of the inherent drawbacks of basic PLCs when applied to safety systems.

Solid-state logic solvers are available to provide high integrity solutions that avoid dependency on software. They are considered particularly suitable for well-defined applications where software flexibility is not required. The need for small and simplistic solutions, however, remains and the option to use simple relay based systems is always available.

5.7 SIS architecture conventions

The following 4 diagrams depict the conventions used for describing some of the more common single and redundant channel architectures including diagnostics. The conventions apply equally to SIS logic solvers, input sensors and output actuators.

These notes are based on material from draft IEC 61511 part 2.

Hardware fault tolerance is the ability of a system to continue to be able to undertake the required safety function in the presence of one or more dangerous faults in hardware. Hence a fault tolerance level of 1 means that a single dangerous fault in the equipment will not prevent the system from performing its safety functions.

From the above it follows that a fault tolerance level of zero implies that the system cannot protect the process if a single dangerous fault occurs in the equipment.

In Figure 5.28

Single channel without diagnostics: 1oo1 has fault tolerance value 0.

Single channel with diagnostics: 1oo1D, still has fault tolerant value 0 but it has a greater safe failure fraction than 1oo1.

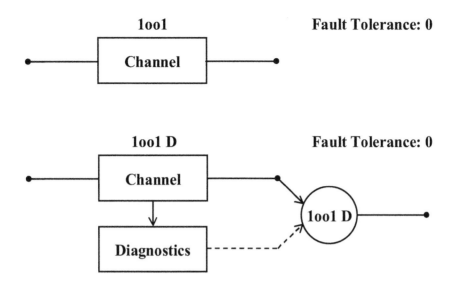

Figure 5.28
Fault tolerance level zero: single channel architectures

In Figure 5.29
Parallel redundant channel without diagnostics: Notation: 1oo2. This design has fault tolerance value 1.

Two channels in series without diagnostics: Notation: 2oo2. This design has fault tolerance value 0

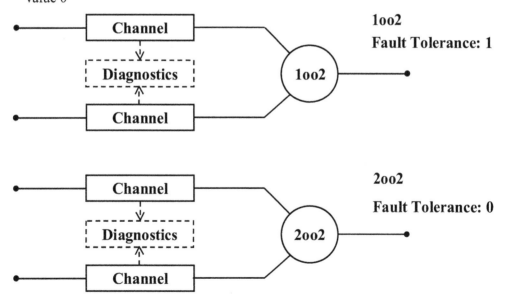

Figure 5.29
Two types of parallel redundant channel

In Figure 5.30
Parallel redundant channel with diagnostics: Notation: 1oo2D. This arrangement uses diagnostics and logic to revert to 1oo1D on detection of a fault in one channel. Fault tolerance level: 1 but with much improved resistance to spurious trips.

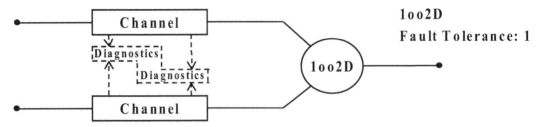

Reverts to 1oo1D on detection of fault

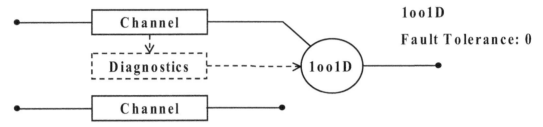

Figure 5.30
A 1002D parallel redundant channel design reverts to 1oo1D on detection of a single fault

In Figure 5.31:

'Two out of three' architecture has three redundant channels with voting arranged to trip if any two channels command a trip. Hence the notation is 2oo3. The fault tolerance level is 1.

If one channel has a dangerous undetected fault the remaining channels must both command a trip before the trip will occur. Under these conditions this architecture becomes the equivalent of 2oo2 and its fault tolerance has degraded to level 0.

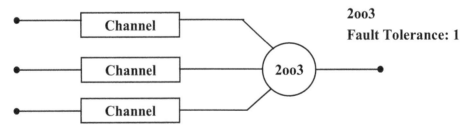

Becomes equivalent to 2oo2 if one channel has a dangerous undetected fault

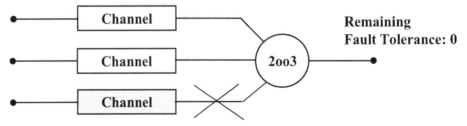

Figure 5.31
2oo3 redundant channel design with majority voting

6

Basic reliability analysis applied to safety systems

6.1   Introduction

In Chapter 5 we started out on the route of conceptual design. We wanted to set out a solution to a given risk reduction requirement in terms of an overall SIS working together with any non-SIS solutions. We didn't get very far along that track because of the need to understand the technology options first. Now that we have done that we must get down to the task of measuring or evaluating the SIS design for its overall safety integrity.

Evaluation of the safety integrity of a design is not something that can be left to intuition or simple estimation. The effects of redundant devices in various configurations cannot be deduced without a reasonable amount of analysis and quantification. In some cases the cost differences between two or three alternative arrangements will be large and the consequences for safety and for financial losses through spurious trips are likely to be even larger.

We need to be able to carry out some basic analysis steps to get us reasonably close to a credible estimate of likely performance. We cannot expect very high accuracy since we are dealing with probabilities and we are using data that is sometimes very limited. But here are several good reasons for doing the best we can with a reliability analysis.

Quantitative analysis:

- Provides an early indication of a system's potential for meeting the design requirements.
- Enables life cycle cost comparisons.
- Determines the weak link in the system and shows the way to improve performance.
- Allows comparisons to be made between alternative designs.
- Supports IEC 61508 requirements for estimating the probability of failure of safety functions due to random hardware failures.

This last point refers specifically to clause 7.4.3.2 of IEC 61508 part 2. It is worth noting here that part 2 of the standard sets out detailed requirements for procedures that are to be followed in the design and development of E/E/PES system. We shall look at these some more when we get to the detailed engineering steps in Chapter 8.

At this stage conceptual design means setting a target value for the probability of failure on demand for the safety function and then estimating the figure likely to be achieved by the proposed solution. If the target is missed the design has to be revised until it is satisfactory.

6.1.1 Design objectives

Before embarking on the trail of reliability analysis it's a good idea to look at what we want to achieve:

Outcome of a reliability study

Analysis of the SIS: What do we want to achieve?

1. Overall SIS probability of failure on demand (PFDavg) for evaluation of risk reduction factor and ability to meet SIL targets.
2. PFDavgs for each section of the SIS (or fail to danger rates if the subject is in high demand rate mode).
3. Effects of different configurations of logic solvers, sensors and actuators on the overall PFDavg. This helps the designer to decide configuration issues.
4. Reliability performance of the instrument sensors and actuators, benefits of redundancy or voting schemes.
5. Guidance/selection of the proof testing intervals.

6.2 Design process

The normal progress of design would bring us to an iterative loop as shown in the next diagram:

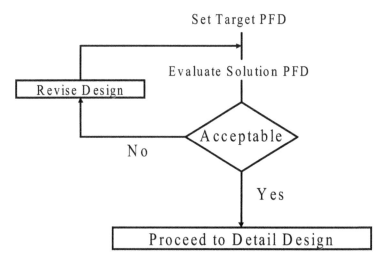

Figure 6.1
Design process

So our first requirement is to develop the PFDavg values for the safety function we are trying to analyze.

First we should revise the terms we are working with.

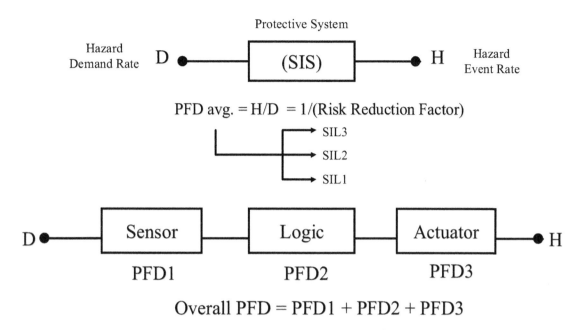

Figure 6.2
Revision of terms

This diagram shows that the demand Rate (D) is the rate (events per year) at which the protective system is called on to act. H is the resulting rate (events per year) at which hazards occur despite the efforts of the protective system.

Hence, as shown before, the risk reduction factor is H/D which is exactly the same as the reduction of frequency of the hazard if no change in the consequence is obtained.

Recalling the term PFDavg, this is the probability that the protective system will fail to operate when required. The PFDavg is also termed 'fractional dead time' (Fdt) from the description that it is the fraction of time that the protective system is inactive. In summary the relationship is simply:

PFDavg = Fdt = hazard rate/demand rate = 1/RRF (for same consequence)

'Ti' is the term used for 'manual test interval'. This period is the time between successive manual tests of a function in the protective system.

'Tia' is the term used for the automatic diagnostic interval used in PES self checking systems.

'MTBF' is the 'mean time between failures' and hence failure rate l is 1/MTBF.

Most of the parameters commonly used in reliability analysis are listed in Chapter 1 and it may be useful to keep this as a reference sheet.

6.3 Failure modes

It is essential to be clear on the failure modes that must be considered for a safety system. The concepts are applicable to all components of the system; instruments, logic solvers and actuators.

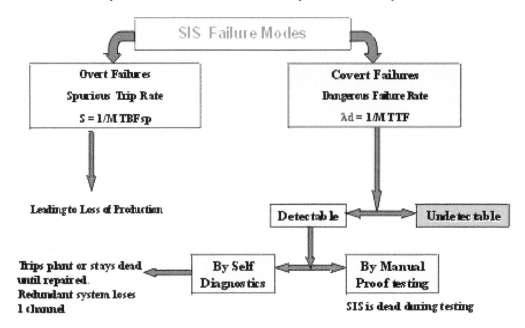

Figure 6.3
Failure modes

The diagram shows there are two basic failure modes, with sub-divisions as follows:

6.3.1   Overt failure mode

Also known as revealed faults because these are faults that become known as soon as they occur. A simple example is a wire break on a sensor connection that is normally carrying signal. Another would be a coil burn-out on a normally closed tripping relay.

Overt failures normally lead to a fail-safe response from the safety system often involving a plant trip. Hence the term 'nuisance trip' is used as well as 'spurious trip'. In a redundant channel SIS an overt failure takes out one channel and the system works on a reduced reliability basis until the fault is repaired. Hence the nuisance trip does not occur until perhaps a second overt failure happens.

6.3.2   Covert failure mode

A covert failure is a dangerous failure until it is detected and rectified. Hence the PFDavg calculation is based on this mode. Typical of covert failures is the stuck relay contact or the PLC output board with a frozen status. These faults will not be self-revealing if the safety system is just dormant with a static logic condition for weeks at a time.

Within the dangerous failure mode there are three sub-categories:

- Faults that can be detected by auto-diagnostics
- Faults that are found by means of periodic manual testing (proof testing)
- Faults that remain hidden undetected in the system until they lead to a failure on demand

Each of the above failure modes has to be considered as a contributor to the PFDavg value for the SIS. Each of the failure modes has a different treatment in the calculation of reliability as we can see in the next section.

The influence of auto-diagnostics on covert failures

It's helpful to be aware at this stage that fast scanning auto-diagnostics can effectively remove the first category of faults from the covert failures list and place them into the overt failures list. This will apply if the SIS is always tripped out (and the plant) when a fault is detected. This would be the case in a single channel system. The alternative approach is to allow the plant to stay in operation without SIS protection until the repair has been completed. In this case the only contribution to failure probability is the portion of time spent repairing the equipment whilst the plant is still running.

In the case of a redundant 'fault tolerant' SIS the offending channel will be out of action until it is repaired but the plant continues to operate with a reduced level of reliability in the SIS until the fault is repaired.

The influence of manual proof testing on covert failures

Manual proof testing reduces the probability of failure on demand because it effectively resets the clock on the cumulative probability of a failure. After a proof test has been successfully carried out the cumulative probability figure is very low and begins to rise with time. Hence the average probability between proof tests is lower than the untested and cumulative probability of failure (which would eventually approach 100%).

6.4   Reliability formulae

This section provides a description of formulae based on simplified versions of well-established reliability analysis methods. The formulae are to be found in many guideline documents and are generally accepted as sufficiently rigorous for the purposes of SIS design. The reliability block diagrams used in this workshop are similar to those defined in IEC 61508 part 6 Annex B where a more comprehensive description of reliability models is to be found.

Material for this section of the workshop was also prepared by reference to Gruhn & Cheddie (ref 1) Chapter 8 where similar formulae will be found.

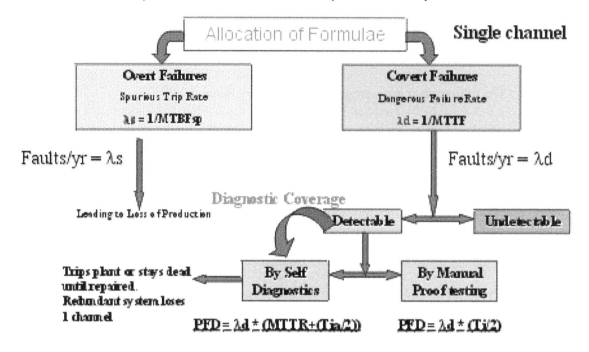

Figure 6.4
Allocation of formulae

The formulae given are applicable to the analysis of the whole SIS or to any subsection. The emphasis is initially on the logic solver with its automatic diagnostics. The manually tested systems we have been talking about are covered by the third set of equations.

In this diagram we can see the basic equations for calculating the nuisance trip rate and the PFDavg values for a single channel component. We shall need appropriate formulae for the same calculations to be performed on redundant or voting configurations. Hence the formulae can be presented in sets to match each of the typical configurations used in SIS design.

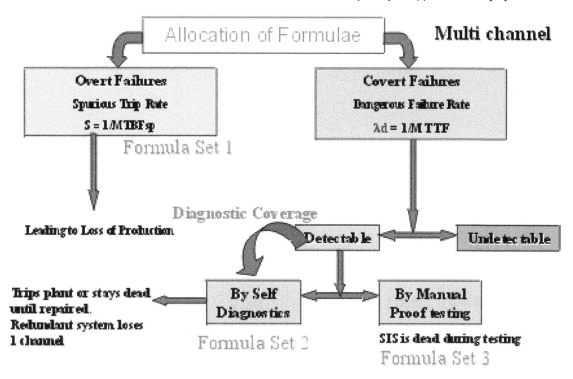

Figure 6.5
Allocation of formulae

Formula set 1 is used to calculate nuisance trips. These will not affect the PFDavg directly but will take the SIS or one of its channels out of service and hence create a loss of availability. The terms 'spurious trip' and 'safe failure' are often used to describe this type of failure and the spurious trip rate is given the character 'S'. MTBFsp is the mean time between spurious trips and for the purpose of practical exercises: λ sp = 1/MTBFsp.

Formula set 2 is used to calculate the PFDavg for any system with automatic diagnostics (or a portion thereof). Note that the test interval, Tia, is very short and should be carried out within the fault tolerant time of the process (*process safety time, hence the need to define this period in the SRS*) if the test itself is not going to affect the availability of the system. Typically Tia would be in the range 1 to 10 seconds.

Formula set 3 is used to calculate the PFDavg for any system with manual proof-testing. Note that the manual proof test interval Ti is to be short compared with the MTBF of the system.

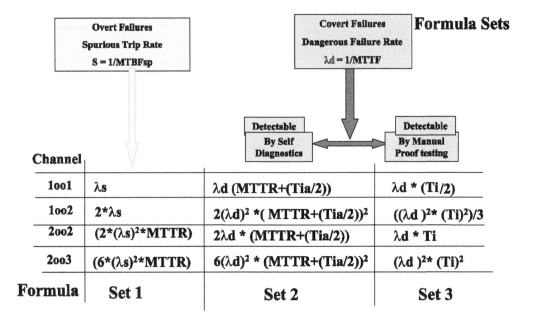

Figure 6.6
Formula sets

6.5 Analysis models and methods

Armed with the above basic formulae we can now set about modeling the SIS into component parts and working out the PFD and spurious trip rates for each section and for the overall SIS function. Here is an example of a basic model with typical values.

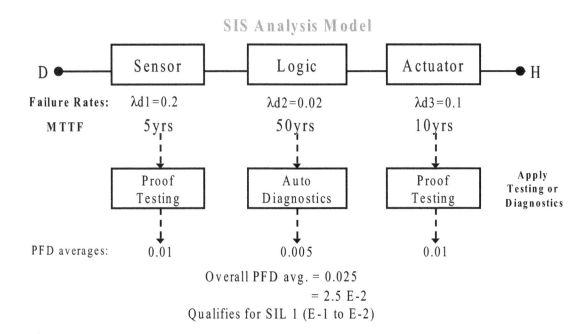

Figure 6.7
SIS analysis model

This diagram shows graphically what we have to do to evaluate the PFDavg of a complete protective system. The example shows how the PFDavgs for each stage of the SIS are set out and summated to give the overall value. The relative values of the different stages allow us to see the weak spots or strong items quite clearly.

6.5.1 Analysis method

Here is a sequence of diagrams that show a step by step procedure for analyzing a typical SIS based on a block diagram representation. Only the dominant failure modes are shown in this procedure to keep it as clear as possible; some refinements will be described later.

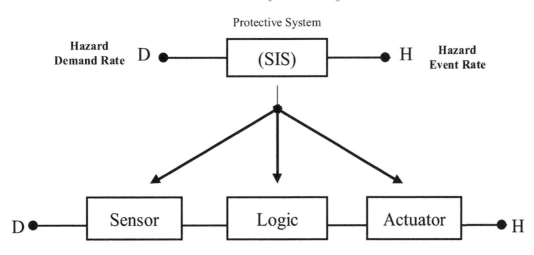

Figure 6.8
SIS analysis: Step 1

The procedure begins by splitting the SIS into the three major stages of sensor, logic solver and actuator. The analysis has to deal with each stage separately before re-combining to establish the overall PFDavg of the SIS.

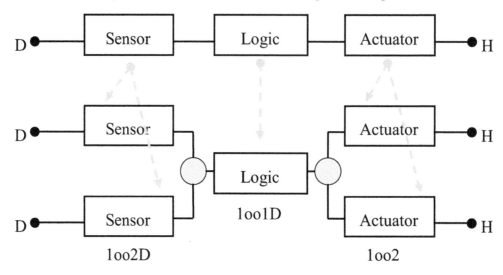

Figure 6.9
SIS analysis: Step 2

In this example the stages are planned to have a redundant pair of sensors with diagnostics allowing one sensor to be disconnected if a fault is found. A single logic solver with diagnostics is used and then a 1oo2 pair of valves or actuators is planned for the final element stage.

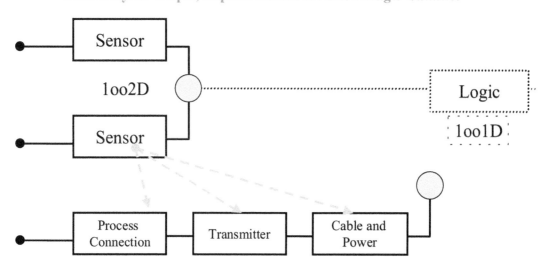

Figure 6.10
SIS analysis: Step 3

Now we need to work out the PFD and spurious trip rates for the single channel of the sensor stage before we can combine the two sensors. To do this we have to break out the components of the sensor channel as shown above in step 3. This now allows us to identify the individual contributors to a possible failure of the single channel sensor loop.

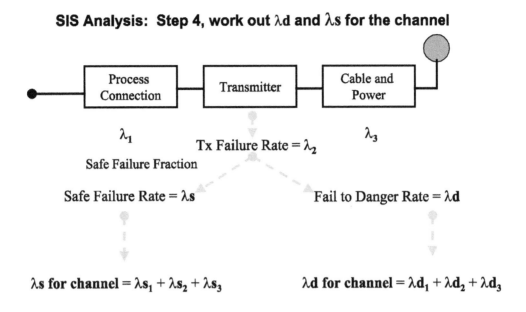

SIS Analysis: Step 4, work out λd and λs for the channel

Figure 6.11
SIS analysis: Step 4

Having found the components we can get down to failure rates for each one. Overall failure rate, l_s, resolves into safe failure, l_d, and fail to danger, l, for each component. The split is determined by the term 'safe failure fraction'. Now we can get the l_s and l_d figure for the whole channel by summing the component failure rates.

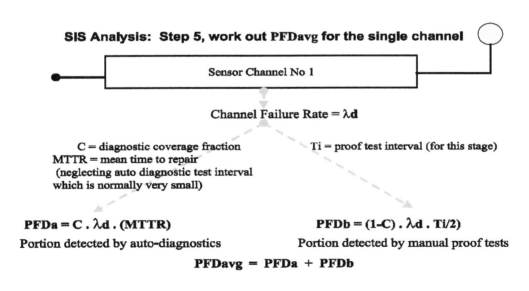

SIS Analysis: Step 5, work out PFDavg for the single channel

Figure 6.12
SIS analysis: Step 5

Now that we have the λ figure for the single channel we can set about the PFDavg. Recalling the allocation of formulae we need to consider the portion of failures that are going to be controlled by automatic diagnostics and the portion of failures that will not be detected until the manual proof test is done. (In our example we ignore the portion that will not be found by either method; we hope this will be negligible but the full analysis method must also allow for this possibility.)

The split between auto-diagnostics and manual proof test is determined by the term C: diagnostic coverage; more on this later. We also need to specify the repair time allowed for the failed sensor. This is MTTR, the mean time to repair.

For the manual proof test portion we need to specify the manual proof test interval, Ti. Because we are using a pair of sensors in this example we do not need to calculate the single channel PFDavg. If it was a single channel the diagram shows how it is done.

Before we move on to the dual channel calculation let's take a look at the way to model the effect of common cause failures.

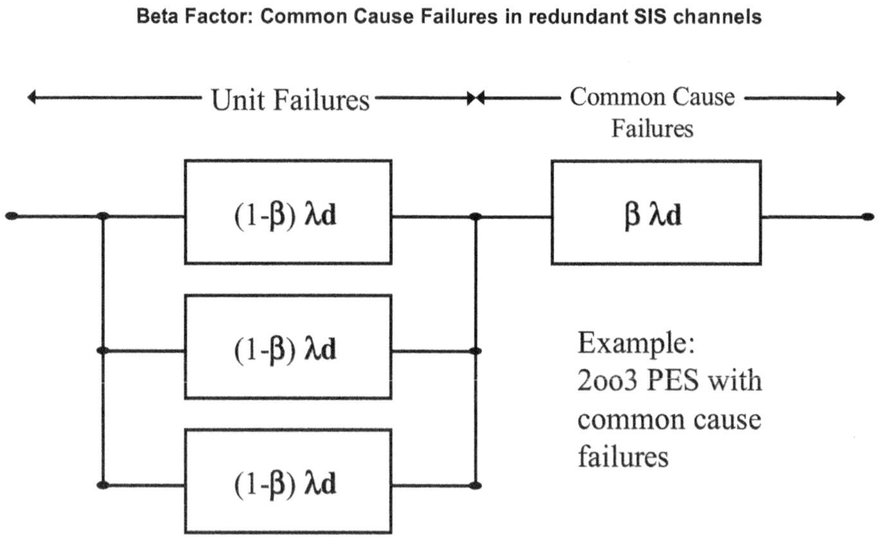

Beta Factor: Common Cause Failures in redundant SIS channels

Figure 6.13
Common cause failures in redundant SIS channels

For conceptual and detail design it is important to be aware that systematic failures due to many factors will limit the minimum values of PFDavg for any part of the SIS that employs redundancy. For example physical deterioration of the hardware of a PES due to harsh environmental conditions may be very significant. It's not a good idea to build a dual redundant PES system into a single cabinet if the cabinet stands beneath a leaking roof!

The general effect of common cause problems is to place an upper limit on the MTBF of any protective system. SIL 3 level systems require diversity as well as redundancy to be considered in order to overcome the limits due to common cause failures.

In the PES and in sensors and actuators the common cause failure fraction is often called the β factor. This is the fraction of failures inherent in the single channel that will still be common to all channels of a redundant system. It should be factored into the reliability calculations as shown in the equations given in the next diagram. It shows how a redundant stage is modeled by adding a single component in series using $\beta \times 1$.

SIS Analysis: Step 6, find the PFD avg for the 1002D sensor group: Break
out the common cause failure fraction for the redundant channels and
calculate PFD avgs for each portion

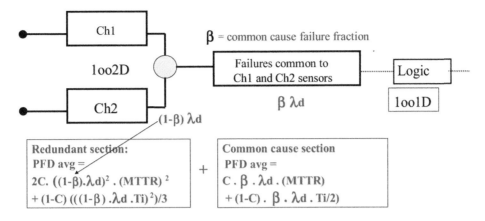

Figure 6.14
SIS analysis: Step 6

Now that we have resolved the factors we need we can proceed to complete the calculation for the redundant sensor pair. The β factor may be typically 10% to 30% for a sensor system without diversity. So we place the common cause failure portion in series with the redundant pair. The contribution to the PFDavg portion for this section is then calculated using formula sets 2 and 3 for the single channel.

The PFDavg figure for the dual section is now obtained by using formula set 2 for the auto-diagnostic portion and formula set 3 for the proof tested portion of failures as shown in the previous slide.

Now we add the two portions of PFDavg to give the PFDavg for the sensor pair.

Example:Dual channel sensors and actuators, single channel logic

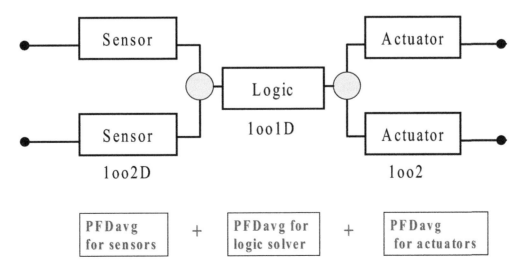

Figure 6.15
SIS analysis: Step 7

The typical procedure that we have now completed is repeated for each section of the SIS. The result is a model as shown in this diagram where we can see the contribution to overall PFDavg made by each of the three stages of the SIS.

Using the results of this model we can obtain the overall SIL value in the IEC table and we see the individual SILs as well. Now if we are using real numbers in our model we should be able to see clearly if the model meets the performance targets of SIL value and required RRF. If it doesn't look good we can see which part is causing the major contribution to the problem.

Example: Dual channel sensors and actuators, single channel logic

Sensor MTTF = 5 years, 75% safe failure fraction. C=0%, β = 10%, Ti = 0.5 yrs, MTTR = 8hrs
Logic MTTF = 10 years, 50% safe failure fraction. C= 95%,β = 10%, Ti = 1 yr
auto diagnostics test interval = 2 secs,   MTTR = 24hrs
Actuator MTTF = 2 years, 80% safe failure fraction. C= 0%, β = 10%, Ti = 0.25 yrs, MTTR = 24hrs

Sensor: single channel λd = 1/5 x .25 = .05/yr   Then PFDavg = .05 x .5/2 = .0125

Logic: single channel λd = 1/10 x .5 = .05
PFDavg for auto test portion = .05 x.95 x 24/8760 = .00013
PFDavg for manual test portion = .05 x.05 x 1/2   = .00125
Total PFD avg for Logic = .0014

Actuator: single channel λd = 1/2 x .2 = .1   Then PFDavg = .1 x .25/2 = .0125

Common cause break out: Sensor = .045 + .005,   Actuator = .09 + .01

SIS Analysis: Example

Example: Dual channel sensors and actuators, single channel logic

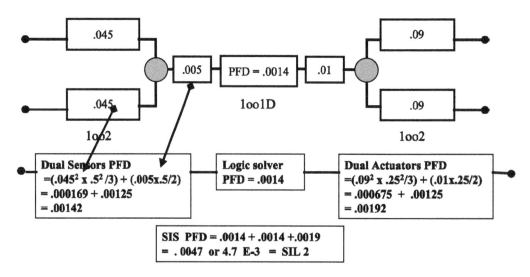

Figure 6.16
SIS analysis: Example SIL calculation

The next diagram takes us through the calculation process with some example data plugged into a model similar to the one we developed. In this case there are no diagnostics in the sensor pair. The derivation of the PFD for the logic solver is shown in the first diagram using the single channel formulae for the fractions with auto-diagnostics and manual diagnostics.

The reliability block diagram shows the fail to danger rates resolved for the individual components and for the common cause fractions where there are redundant pairs. The formula for 1oo2 voting (manual proof testing only) is used for the redundant pairs and the single channel formula is used for the common cause and single channel logic components.

The result shows that the total PFDavg is: 0.0047 and the overall integrity is SIL 2.

The percentage contribution to failures is as follows: Sensor pair: 30%, Logic: 30%, actuator pair: 40%. From this we can see that this design is reasonably well balanced for a SIL 2 objective but would need substantially better reliability figures to meet SIL3.

Remember that this is the model for hardware failures and it does not include the potential systematic failures or associated software failures. This highlights the benefits of the new approach to engineering of the system design and the engineering of the software because these factors cannot be built into a quantitative hardware analysis.

6.5.2   Calculations for spurious trips

The next step in our analysis activities will often be to check the position regarding spurious trips. The design may meet the SIL targets but it may be at the price of a high rate of lost production through unwanted trips.

SIS Analysis: Example Calculation for Spurious Trip
Example: Dual channel sensors and actuators, single channel logic

Sensor MTTF = 5 years, 75% safe failure fraction. C=0%, β = 10%, Ti = 0.5 yrs, MTTR = 8hrs
Logic MTTF = 10 years, 50% safe failure fraction. C= 95%,β = 10%, Ti = 1 yr
auto diagnostics test interval = 2 secs,   MTTR = 24hrs
Actuator MTTF = 2 years, 80% safe failure fraction. C= 0%, β = 10%, Ti = 0.25 yrs, MTTR = 24hrs

Sensor: single channel λs = 1/5 x .75 =.15/yr

Logic: single channelλs = 1/10 x .5 = .05

Actuator: single channelλs = 1/2 x .8 = .4

Common cause break out: Sensor = .135 + .015,   Actuator = .36 + .04

SIS Analysis: Example, Spurious Trip Rate

Example: Dual channel sensors and actuators, single channel logic

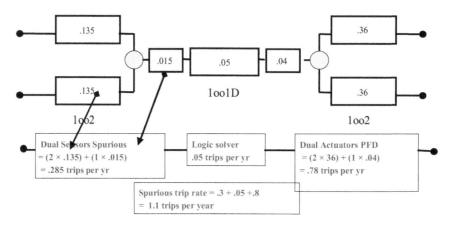

Dual Sensors Spurious
= (2 × .135) + (1 × .015)
= .285 trips per yr

Logic solver
.05 trips per yr

Dual Actuators PFD
= (2 × 36) + (1 × .04)
= .78 trips per yr

Spurious trip rate = .3 + .05 +.8
= 1.1 trips per year

Figure 6.17
Calculation for spurious trips

The first diagram shows the calculation of single channel l's values. These values are then placed into the reliability model diagram. The calculations use formula set 1 to derive the component spurious trip rates. These rates operate in series in the model so the rates can be summed to give a net spurious trip rate of 1.1 trips per year.

Depending on the nature of the process this may or may not be acceptable. For a process with high downtime costs the pressure would be on to improve this figure. The next diagram shows a trial improvement in the design.

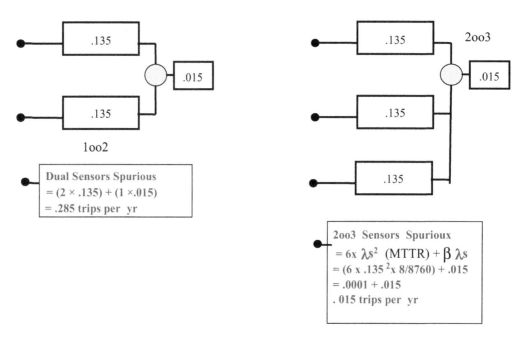

Dual Sensors Spurious
= (2 × .135) + (1 ×.015)
= .285 trips per yr

2oo3 Sensors Spurioux
= 6x λs^2 (MTTR) + β λs
= (6 x $.135^2$x 8/8760) + .015
= .0001 + .015
. 015 trips per yr

Figure 6.18
Reducing spurious trip rate

In this example the sensor pair has been changed to a 2oo3 voting group. The calculation again uses formula set 3 with the appropriate choice of formula. The result shows a much improved spurious trip rate for the sensor section, as we would expect. However when adding in the contributions from the rest of the SIS loop we see the net rate has only improved to 0.84 trips/yr. It would have been better to go for 2oo3 voting on actuators or search for a more reliable actuator.

6.5.3   Conclusions on analysis models

Reliability models are reasonably straightforward and are very helpful for visualization of the performance factors of the SIS. The problem is that the calculations can be tedious and prone to errors. The quality of input data is also a limiting factor in accuracy of results.

However there is much that can be done to assist with this problem. Firstly the calculations lend themselves to computer modeling and calculation packages so that as long as we understand the principles software tools can make the manipulation of the data much easier.

IEC 61508 part 6 provides details of reliability analysis models similar to the example given in this chapter and provides more precise details of the failure modes that should be factored into the equations. The standard then provides a convenient set of look up tables for typical ranges of data such as λ and β factors. If you become familiar with the use of these tables it represents an alternative to direct calculation. The advantage of this approach is that it references published data and is therefore easy to validate. This will be particularly useful for developers of calculation packages since they can validate the models against the IEC data tables.

Finally we should note here that fault tree analysis methods provide an alternative to reliability block diagrams, particularly for detailed failure rate analysis of sub systems. Hence electronic and PES manufacturers will use these methods to evaluate the reliability of their specific products.

When dealing with complex sub systems such as 1oo2D logic solvers the reliability modeling technique known as Markov modeling has become established as one of the best means of depicting failure modes of complex systems. We need to be aware of this technique because of its relevance to calculation methods for redundant channel systems and because it is extensively used by vendors to support the performance figures for logic solver reliabilities. We shall briefly discuss the elementary principles of this technique but its application is outside of the scope of this book.

For the conceptual design stage we have the choice of high-level fault trees or reliability block diagrams.

6.6   Some design considerations

Continuing with our evaluation of the SIS design here are some additional factors we should keep in mind during the analysis work.

6.6.1   Proof testing basics

One of the contributors to improving the PFDavg will be the application of manual proof testing. We have some control over this factor since we can adjust the proof test interval to improve the PFDavg values for those items in that category.

The formula relating PFDavg to the fail to danger rate of the single channel system under consideration is derived as follows:

The cumulative probability of failure of a single channel device is $p(t) = 1 - e^{\lambda d.t}$
This is shown on the next diagram.

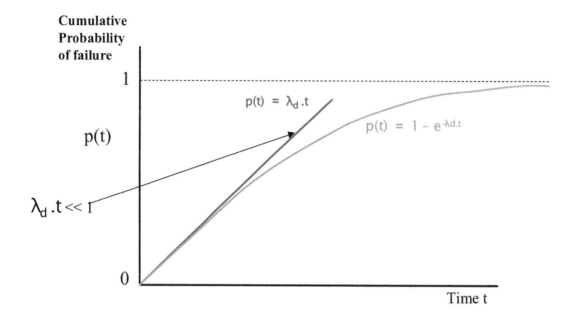

Figure 6.19
Cumulative probability of failure

Normally $l_d.t$ is much less than 1 so that $p(t) = l_d.t$ and a straight line relationship exists between p and t with slope $= l_d$.

Now the PFDavg is defined as the average failure rate on demand over a period t. When manual proof testing takes place in an interval Ti such that $l_d.t$ is much less than 1 the PFDavg is given by

PFDavg $= 1/\text{Ti} \int_0^{\text{Ti}} (l_d.t) \, dt$

which gives

PFDavg $= l_d.\text{Ti}/2$

This basic relationship is used throughout the simplified reliability calculations for the reliability of a single channel device. The significance of this expression is that PFDavg can be made smaller than l_d by proof testing.

In order to arrive at a hazard rate, H, it is simply a matter of multiplying the demand rate for the protection system to operate, D, by the average probability that it will fail at the time; i.e. the PFDavg.

$H = D.$ PFDavg

or for a trip test period Ti:

$H = l_d .D.$ Ti/2

This relationship is only true for the simplified condition given above. A more accurate version is shown in the next diagram where hazard rate, H, is shown plotted against demand rate D.

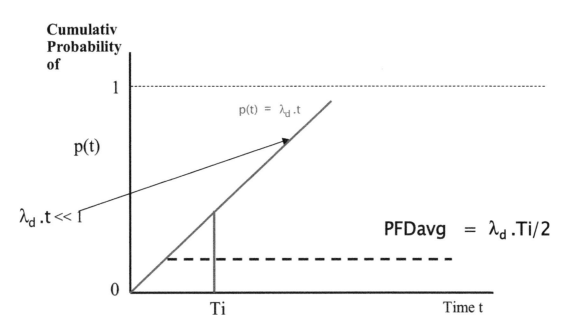

Figure 6.20
Cumulative probability of failure

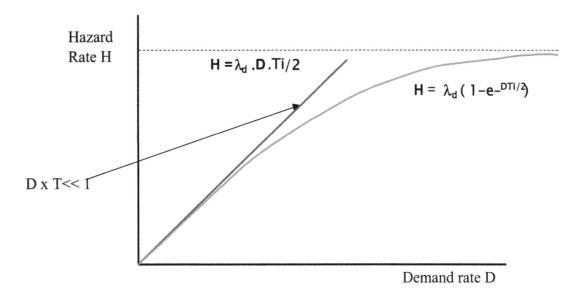

Figure 6.21
Hazard rate vs demand rate

The relationship between hazard rate and demand rate is shown for an accurate version based on:

$$H = l_d(1 - e^{-DTi/2})$$

and a simplified version:

$$H = l_d.D.Ti/2$$

where the term $D.Ti$ is much less than 1.

Cautionary note 1

The above diagram shows the difference between the two versions of the above equations for calculating hazard rate which means that the simplified version yields pessimistic results when the demand rate is high or the test interval is a long one.

However the straight-line approximation is a good fit to most realistic situations and this simple equation is used as the basis for the evaluation of most applications where the demand rate is low.

Cautionary note 2

The PFDavg equation demonstrates the simple principle that the original failure rate λd of the non-diagnostic portion of the SIS can be substantially improved by periodic proof testing. This is the basic reason for doing the tests.

The temptation is to drive down the PFDavg figure by increasing the proof test frequency. However this approach has its limitations as we can see in the next paragraph.

Limitations of manual proof testing

The formula given for manual proof testing above must be corrected for the effects of downtime whilst the test is in progress and for the fact that there may be errors in the testing or that the test work may itself leave the system in a less than healthy state. For example the calibration of a sensor may have been wrongly done. Then the equation becomes:

PFDavg = $(D.\text{Ti. ld}/2) + (\text{MTD/Ti}) + E$

where E is a constant for the probability of errors due to testing and MTD is the manual test duration.

This leads to the interesting graph in the next diagram where the PFDavg is seen to reach a minimum value as Ti is reduced but rises sharply again as Ti is reduced further.

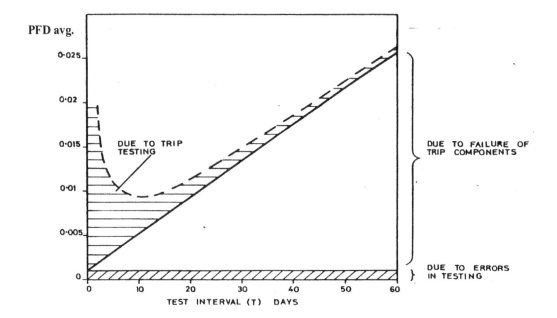

Figure 6.22
Limitations of manual proof testing

6.6.2 Reliability in a high demand mode

The limitations and approximations given above are presumed to be the reasons why IEC 61508 will not allow the PFDavg value to be used as the measure of SIL if the demand rate is high. If the demand rate D is high and if the test interval cannot be short then DTi becomes large.

In this case the term $e^{DTi/2}$ in the hazard rate equation, $H = l_d(1 - e^{DTi/2})$ trends towards zero.

Then: H approaches l_d and proof testing is of no value.

So for high demand rate applications the reliability of the overall system must be expressed as the simple fail to danger rate; i.e. we can assume the demand is continuous and dangerous failures occur as soon as the equipment fails.

6.6.3 Comparison of protective systems

Here is a useful table for the comparison of protective systems based on the formulae we have been using.

System	Fail Safe Fault Rate Faults/Year	Fail to Danger Fault Rate Faults/Year	Fractional Dead Time	
1 out of 1	S	F	$\frac{1}{2}$ FT	= fdt
1 out of 2	2S	F^2T	$\frac{1}{3}F^2T^2$	= 4/3(fdt)^2
2 out of 2	S^2T	2F	FT	= 2 fdt
1 out of 3	3S	F^3T^2	$\frac{1}{4}F^3T^3$	= 2(fdt)^3
2 out of 3	$3S^2T$	$3F^2T$	F^2T^2	= 4(fdt)^2

The above table assumes that the subsystems are identical.

The following general equation can be used to obtain the above values –

$$\text{fdt (m out of n)} = {}^{n}C_r \frac{F^r T^r}{(r+1)}$$

$$\text{where} \quad r = n - m + 1$$

$$^{n}C_r = \frac{n!}{r!(n-r)!}$$

$$\text{and } n! = n \times (n-1) \times (n-2)x \ldots\ldots x3x2x1$$

Figure 6.23
Comparison of protective systems

These formulae only apply to systems with manual testing and are the same as our formula set 3. They show for example that with a relay type system where manual testing is the only form of diagnostic the use of a 2oo2 system is inferior to a 1oo1 system.

E.g. for a relay system with a fault rate of 0.2 dangerous faults per year and a test interval of 0.25 years:

PFDavg = ½ × 0.2 × 0.25 = 0.025

For a 2oo2 dual channel version of the same system:

PFDavg = 0.05

But for a 2oo3 voting system:

$$PFDavg = \frac{1}{4} F^3 T^3 \text{ or } 2(fdt)^3 = 2 \times .025^3$$
$$= 0.00003 \text{ failures/yr}$$

These PFD figures do not allow for common cause failures or any other factors. These must be taken into account and the correct analysis procedures must be deployed to arrive at a dependable answer. For the manufacturers of SIS products and for certification authorities there is considerable specialist skill required in this field. We can only touch on some areas to give an idea of what is involved.

6.6.4 Markov models

A brief diversion here to illustrate what is meant by Markov models, also known as failure state diagrams.

Markov analysis is a reliability and safety modeling technique that uses state diagrams. The technique uses two symbols: a circle and an arc. A circle represents a typical system state that is characterized by a unique combination of component failures. An arc represents the probability to go from one state to another due to component failure or due to component repair.

This diagram shows an example of a Markov model.

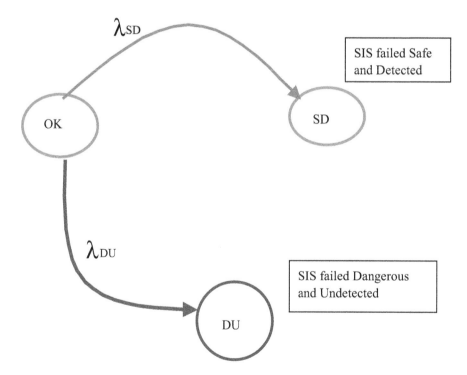

Figure 6.24
Markov model

The models can be used to represent all the potential states of the equipment and to show all possible routes for entry to and exit from the states.

In the case of reliability analysis for an SIS the following typical failure modes are used.

Each component of a safety-related system can fail in four different ways:

- fail to safe and the failure is detected (state code: SD)
- fail to safe and the failure is undetected (state code: SU)
- fail to dangerous and the failure is detected (state code: DD)
- fail to dangerous and the failure is undetected (state code: DD)

For the SIS and its PES repairs can be carried out and are added into the model as follows:

The repair rate, μR is defined as the probability of a failure being repaired per hour. The repair rate for detected failures is calculated as 1/MTTR. Probabilities of each state are calculated as a function of time.

The next diagram shows this version with the added detail of proof testing of the SIS. In this case, the PFDavg may be calculated by considering the ratio of time in state DU to the time in the OK state.

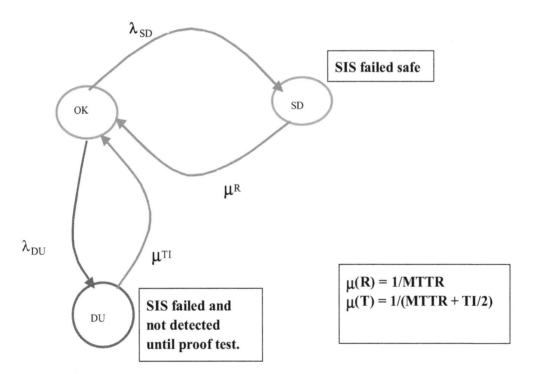

Figure 6.25
Proof testing on the SIS

The fraction of any operating period (PT) spent in state SD is:

$PT \times \lambda_{DU}(MTTR + TI/2)$

PFDavg is (time in SD)/PT so PFD avg $= \lambda_{DU}(MTTR + TI/2)$

For an example of the application of this method of modeling, look at the technical paper published by Moore Instruments and written by Bukowski and Goble for the ISA 1994 conference. It provides a clear illustration of how Markov models are used to evaluate the reliability characteristics of different redundant configurations.

6.6.5   Diagnostic coverage

As we know from the previous chapter the use of automatic diagnostics is essential for obtaining a satisfactorily low rate of fail to danger for a PES. The diagnostics convert a proportion of the potentially dangerous undetected faults into fail to safety faults. The diagnostic coverage factor is therefore the percentage of dangerous failures that the system will detect automatically.

A most effective illustration of this uses a simple Markov model as seen in the next diagram.

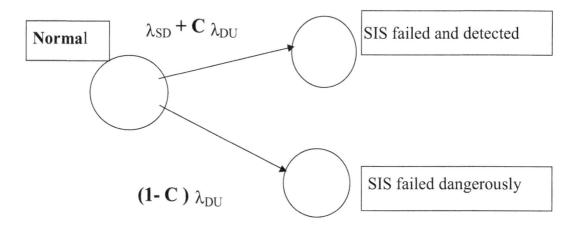

Figure 6.26
Diagnostic coverage

This material references part of a useful paper on reliability analysis by: Steven E. Smith: Fault coverage in plant protection systems, published in ISA technical papers, Volume 21, 1990.

As diagnostic coverage C increases the MTTF(D) increases dramatically whilst the MTTF(S) only falls by 50% (assuming $\lambda_{DU} = \lambda_{SD}$). The next diagram illustrates the effect of increasing diagnostic coverage.

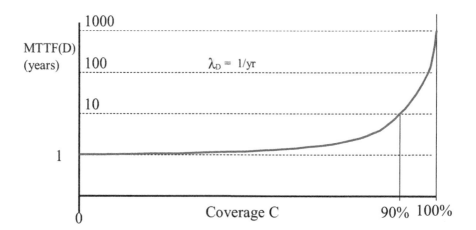

Figure 6.27
Effect of increasing diagnostic coverage

Note that DC factors below 80% do not help very much. At 90% there is a real benefit and any improvement beyond this value yields dramatic improvements in MTTF (D). Manufacturers will obviously strive to get their DC figures above 95% and if possible up to better than 99%.

6.6.6 Reliability calculation software tools

We have seen here some of the basic techniques needed to evaluate the reliability of an SIS configuration. Although the focus has been on the PES logic solver the main task is still in analyzing the overall configuration. Spreadsheet software is the obvious tool to use if the calculations are frequently used and are of a similar format each time they are needed. A good example of a practical tool kit is the Honeywell Safecalc package. The package allows the overall SIS to be analyzed in wide combination of configurations and by entering appropriate values for failure rates or MTTRs the user can arrive at overall PFDavg values. The SIL level is automatically given and any basic constraints on configuration are stated.

6.6.7 Summary

Here is a summary of the relevance of reliability analysis in support of the conceptual design phase.

Intuition is not as good as quantitative analysis.

Analysis provides early indication of potential to meet SRS:

- Exposes weak links
- Failure modes
- Nuisance trip rate
- Dangerous failure rate

Analysis leads to PFDavg, leads to SIL, SIL matches the SRS.

6.7 Summary of parameters used in the reliability analysis of the safety systems

Abbreviation	Parameter	Explanation
MTTF	Mean Time to Failure	Mean period in years or hours for a component to fail. Same as MTBF. Statistical mean of data, not the life of a component.
MTTR	Mean Time to Repair	Mean period in hours for offline repair of component.
PFD avg	Probability of Failure on Demand	Average probability of system failing to respond to a demand.
H	Hazard Rate	Rate in events/yr or hr at which hazards occur as result of failure to protect against demands.
D	Demand Rate	Rate in events/yr or hr at which demands are placed on the safety system to operate.
Ti	Proof Test Interval	Must be in hrs or yrs to be consistent with H and D.
Tia	Auto Diagnostic Period	Interval between cycles of self testing in a PLC or instrument.
λ	Failure Rate	Rate in failures per hour for the component
λd	Fail to Danger Rate	Generally suffixes are used to denote the various failure modes. Hence: D = danger DD = danger detected DU = Danger undetected
λs	Fail to Safe Rate	The rate of failure to a safe mode (tripped) Hence: D = safe SD = safe detected SU = safe undetected
C	Coverage	Diagnostic coverage converts potentially dangerous failures into detected failures through diagnostic testing. High C factors reduce the PFD avg of a PES.
β	Common cause factor	Proportion of potentially faults that are common to a multi-channel protection system.

6.8   Some sources of reliability data for instrumentation

Failure rates for instrumentation used in safety instrumented systems should be calculated as a sum of the possible failures of the process connection, the sensor or transmitter itself and its associated cabling and power supplies. Similarly failure rates for actuators must take into account air supplies, solenoids, drive cylinders and valves.

Often the data published by instrument vendors applies only to the device itself and excludes the other factors listed above. Hence responsibility lies with the end user to develop the overall failure rate for the application.

Sources of data

The following sources of failure data have been compiled by Simmons Associates, Engineering and Software Management Consultants. Simmons Associate publishes a series of downloadable 'Technical Briefs' in safety-instrumentation and software quality management. These can be found on their website: http://www.tony-s.co.uk.

Reliability data in books

- *Guidelines for Safe Automation of Chemical Processes*, 1993 New York: American Institute of Chemical Engineers Pub. No. G-12. ISBN 0 8169 0554 1, $120
- *Reliability Maintainability and Risk – David J Smith*, Butterworth Heinemann. 6th ed, Feb 2001 ISBN 0 7506 5168 7.
- OREDA (Offshore Reliability data) book 1997 – DNV Industry Norway: Offshore
- Reliability Data Handbook 3rd Edition, 1997 (OREDA-97)

OREDA (Offshore REliability DAta) is a project organization with eleven oil companies: AGIP, Amoco, BP, Elf Petroleum, Esso/Exxon, Norsk Hydro, Phillips Petroleum, Saga Petroleum, Shell, Statoil and TOTAL. OREDA's main objective is to provide a basis for improving safety and cost-effectiveness in the offshore oil and gas industry by the collection and analysis of reliability data, and promoting the use and exchange of reliability know-how among the participating companies.

See http://www.sintef.no/sipaa/prosjekt/oreda/handbook.html
OREDA Publications are obtained via SINTEF – see below.

- Center for Chemical Process Safety (CCPS) in its Process Equipment Reliability Database
- Project (PERD). see   http://www.aiche.org/ccps/perd.htm
- Reliability Prediction of Electronic Equipment – MIL-Handbook – 217. See http://www.reliabilityprediction.com/milstandard/milhandbook.htm

Currently there are 9 MIL Handbooks available; the documents include the original MIL document with any applicable revisions and releases. The MIL Handbooks available on CD are:

- MIL-HDBK-189 Reliability Growth Management
- MIL-HDBK-217 Reliability Prediction of Electronic Equipment
- MIL-HDBK-251 Reliability Design Thermal Applications
- MIL-HDBK-338 Electronic Reliability Design Handbook

- MIL-HDBK-344 Environmental Stress Screening (ESS) of Electronic Equipment
- MIL-HDBK-472 Maintainability Prediction
- MIL-HDBK-502 Acquisition Logistics
- MIL-HDBK-2084 Maintainability of Avionic and Electronic Systems and Equipment
- MIL-HDBK-2164 Environmental Stress Screening Process for Electronic Equipment
- SINTEF Industrial Management Safety and Reliability

SINTEF Industrial Management, Safety and Reliability performs contract research within the safety, reliability, maintenance and quality disciplines. The department has extensive experience in supporting the petroleum sector, onshore industry, transport (air, maritime and rail) and the public sector with solving problems within its work areas. Practical problem-solving skills combined with scientific competence and analytical capabilities have resulted in several innovations in safety and reliability techniques
See http://www.sintef.no

- IEEE Standard 500: 1984 – still referred to as a source of reliability data but now obsolete.

CONTACT DETAILS

Butterworth Heinemann
Butterworth-Heinemann
225 Wildwood Avenue, Woburn, MA 01801, USA
Phone: 800-366-2665 Fax: 781-904-2620
E-mail custserv@bhusa.com
Butterworth-Heinemann
Linacre House, Jordan Hill, Oxford OX2 8DP, UK
Phone +44 (0)1865 310366
See http://www.bh.com

Center for Chemical Process Safety (CCPS)
Center for Chemical Process Safety
American Institute of Chemical Engineers
3 Park Ave, New York, N.Y., 10016-5991, U.S.A.
Tel. Toll Free: 1-800-AIChemE, (1-800-242-4363),
Tel: (212) 591-7319 Fax: (212) 591-8895
E-mail: ccps@aiche.org
See http://www.aiche.org/ccps/perd.htm

DNV
Det Norske Veritas
RN570,Veritasweien 1, N-1322 Norway
Telephone: +47 67 57 83 91 Fax: +47 67 57 99 11 Attn: RN570
E-mail: oreda@dnv.com
See http://www.dnv.no
For OREDA see http://www.dnv.no/environmentalcons/default.htm

IEEE
Institute of Electrical and Electronic Engineers
Standards Information
Tel: 1-732-562-3800
Fax: 1-732-562-1571
E-mail: stds-info@ieee.org
IEEE European Operations Center (Brussels)
Tel: 32-2-770-2242
Fax: 32-2-770-8505
E-mail: memservice-europe@ieee.org

6.9 Safety performance calculation packages and reliability databases

Calculation packages for evaluating the SIL requirements and SIL ratings of safety systems are becoming available from several companies. In particular, the following companies offer SIL calculation packages with options to use integral generic and/or instrument-specific failure rate data.

- Asset Integrity Management Ltd: Aberdeen, Scotland.
  See: http://www.assetintegrity.co.uk

SILClass Application Tool provides:

- SIL determination
- Reliability data
- Design architecture selection
- Maintenance cycle optimization
- PFD and safe failure fractions

Exida.com L.L.C. Sellersville PA USA and München, Germany

See: http://www.exida.com

- SILVER – SIF quantitative verification tool. Calculates SIL and PFD values for detailed instrumented safety functions. Contains integral reliability data for commonly used instruments.

7

Safety in field instruments and devices

7.1   Introduction

We have reached the stage in the safety life cycle where we need to be sure of getting things right in the design of the field instrumentation to be used in any planned SIS.

This chapter looks at a range of instrumentation design techniques that have accumulated in the industry through experience which began a long time before the days of PES and the high performance logic solvers. In this area we find well established design rules but the introduction of intelligent instruments and field bus type communications presents new challenges and some pitfalls for the design team.

The subjects to be covered are:

- The impact of field devices on safety integrity
- Potential failure modes and their causes
- Issues of separation, redundancy and diversity
- Fail-safe design basics
- Development of diagnostics
- Safety requirements in the selection of instruments
- Safety certified transmitters and actuators
- Smart devices and bus technology

The material in this book is based on the process instrumentation point of view. The principles laid down for the process control industry appear to extend pretty well into any automated plant or machinery system. The difference in machinery systems is one of speed and perhaps more specialized sensing. Issues of diagnostics, redundancy and installation are all common to many industries.

7.2   Objectives

- To be aware of the key features of field instruments required to support a good SIS
- To be able to specify the basic instrument arrangement most suited to the needs of a particular SIS
- To be able to offer preliminary solutions to shutdown system problems at the hazard study stage
- To know what the standards require for instruments.

Field instrument selection and installation design is a subject where experience provides the best guide; but many companies do not have experienced instrument engineers. In practice many process engineers or mechanical engineers need to have some knowledge of instrumentation to perform their role in the safety system design. Some additional sources of guidance are listed here:

- Appendix B of ISA S84.01 has basic guidelines for instruments in safety applications.
- The ISA guidelines have largely been transferred into Annex B of the upcoming IEC 61511 part 2.
- IEC 61508 part 2 has specific requirements for the selection and installation of safety-related instruments. IEC 61511 part 2 has expanded these specifically for process industries.
- Gruhn & Cheddie has basic design guidelines and many helpful points.
- Instrumentation buyers guides provide a quick guide to available products.
- Vendors. The larger instrument companies publish application guides.

7.3   Field devices for safety

Field devices basically comprise the sensors to provide the input information to the logic solver and the actuation devices required to carry out the trip function when the demand comes along. They are supported by the wiring and process connection arrangements; taken together they comprise the area with the greatest potential for problems. They are also the area where most engineers have a chance to exercise their skills and judgment in design, selection and maintenance. You cannot just 'buy in' a solution to this problem.

7.3.1   Key points about sensors and actuators

Some important points to keep in mind when dealing with sensors and actuators are summarized. The next paragraphs will consider these and other issues.

- Sensors and actuators remain the most critical reliability items in an SIS.
- Separation, diversity and redundancy are critical issues.
- Safety related instruments must have a proven record of performance. IEC 61508/61511 have specific requirements.
- Logic solver intelligence and communications power will help to provide diagnostic capabilities to assist field device reliability.
- Failure modes and common cause issues are potential problems for intelligent instruments.

7.3.2   Sensors and actuators dominate reliability issues

Consider the 1st point in the above list: A typical reliability table will illustrate why the field devices are the major contributors to the possible failures of most safety functions.

Item	Fail to Danger Rate/yr	PFDavg (3 month proof test)	PFDavg % of total
Input sensor loop	0.05	0.006	32
SIL 2 Logic Solver PLC		0.0005	3
Output Actuator loop (solenoid + valve)	0.1	0.0125	65
Totals		0.019 (Sil 1)	100

Table 7.1
Typical reliability data shows how field devices dominate SIS failure rates

The field devices taken together contribute 97% of the PFD for this example. Note also that the PFD figures for the field devices are affected by environmental conditions and maintenance factors such as test procedures, calibration methods and the effects of mechanical plant maintenance activities.

All of this means that once an adequate logic solver has been chosen for the SIS the bulk of the engineering and maintenance effort should concentrate on the field devices. This fact is sometimes masked by all the effort that has gone into improving the performance of logic solvers. Possibly this is because the field systems are application specific rather than product specific as is the case with PLCs; i.e. vendors can improve the PLCs as products but end-users have to look after the field instruments as applications.

Note that the PES logic solvers include the benefits of auto-diagnostics to improve the PFD figures. It is more difficult to do the same for sensors and actuators but in cases where this can be done there will be major improvements to the PFDs.

7.4   Sensor types

What do we mean by sensors? We mean any device that is able to represent the value or status of a chosen parameter in a form suitable for decision making in the SIS logic solver. Some illustrations are shown in the next diagrams.

There are only two basic categories of sensors: switches or transmitters.

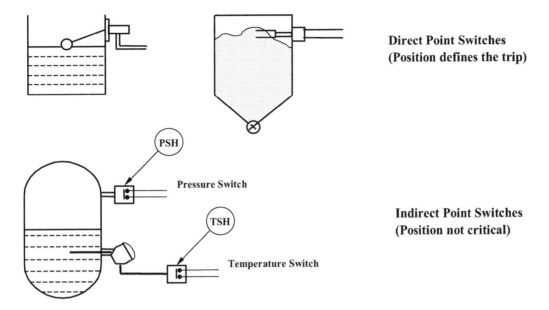

Direct Point Switches
(Position defines the trip)

Indirect Point Switches
(Position not critical)

Figure 7.1
Categories of sensors

- Switches:
  Simple electrical or pneumatic switches that produce a state change when the limiting condition is exceeded. These can be subdivided into sub classes:
  - Direct point switches where the sensor's physical location defines the limit value; e.g. limit switch or proximity switch; a fixed level switch such as the float switch commonly used on small boiler drums.

  - Indirect point switches such as a pressure switch or temperature switch where the basic measurement is converted mechanically or electrically to a signal which has an adjustable limit value.

Transmitters

These employ sensing elements where a signal representing a process parameter is continuously sent to the logic solver. The limiting condition is then determined by the logic solver rather than the sensor. These can be subdivided into some typical variants as shown below:

- 4–20 mA current transmission to an input A/D converter stage in the logic solver. The field instrument is powered by a portion of the loop current supply.
- 4–20 mA current transmission or a voltage signal typically 1–5 V transmitted from a separately powered converter stage either control room or field mounted.
- Low level signal (from thermocouple or RTD element) directly to an input converter stage in the logic solver.
- Digital signal transmission using a bus protocol such as field bus or profibus from a field mounted transmitter to a bus converter stage at the logic solver.

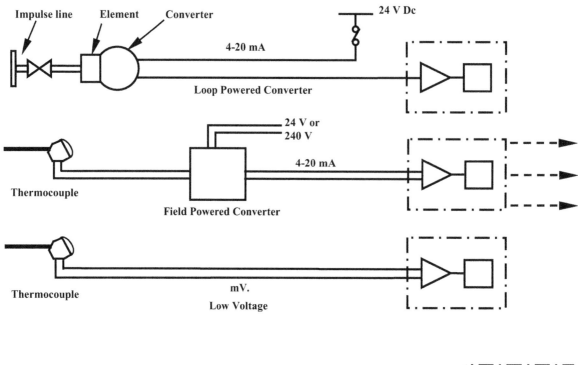

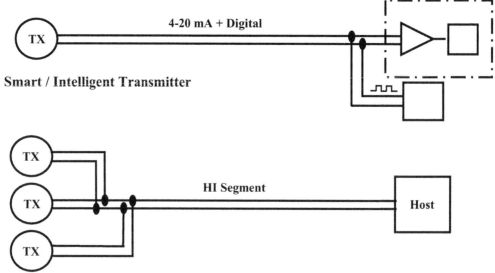

Figure 7.2
Typical communication arrangements for transmitters

7.4.1   Using transmitters with trip amplifiers

Panel mounted trip amplifiers or field mounted current/voltage alarm trip devices have
been used for many years in safety systems to convert analog signals from transmitters to

logic state signals defining healthy or tripped conditions. The logic signal, usually as a relay contact pair, is then used either as an input to a logic solver or is sometimes used directly to drive a tripping output device such as a motor starter or solenoid valve.

Panel based trip amplifiers are designed to have a good safe failure fraction and are well accepted for use with relay based trip systems. Trip amplifiers are normally specially designed to have a high level of immunity to RFI disturbances and have programmable responses to sensor failures and to some internal failures.

Trip amplifiers have the disadvantage of adding another component to the safety loop with potentially dangerous failure modes. Programmable logic solvers employ integral analog-to-digital conversion of the transmitter signal and hence eliminate the need for a separate device to detect trip levels.

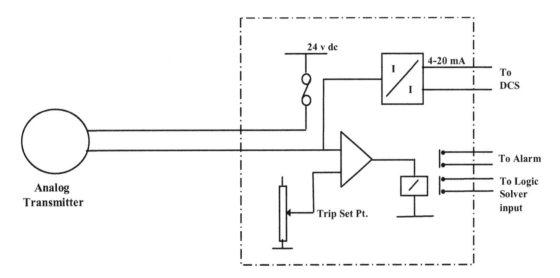

Figure 7.3
Trip amplifier for interfacing an analog transmitter into the logic solver

This diagram shows a typical trip amplifier arrangement. Note how retransmission facilities are provided for the analog signal. This allows the signal to be copied to a control room interface or to a DCS without adding complexity (and hence failure potential) to the input signal loop.

However the use of simple current/voltage alarm contacts from a converter unit as tripping devices should be avoided wherever possible. These instruments are likely to have unpredictable failure modes and their field adjustable trip settings are difficult to protect against unauthorized or temporary changes.

Advantages of analog transmitters over switches

In concluding this review of sensor types we must note that all the standards for safety systems advocate the use of analog transmitters in preference to switches or sensors for safety applications. The advantages are listed in the above diagram and should be noted carefully when planning a safety system:

- Good reliability and accuracy
- Signal present at all times
- Potential for diagnostics, easier to detect faults
- Possible to compare signal with other parameters

- Trending and alarming available
- Multiple set points
- Competitive pricing
- Rationalized spares

7.4.2   A list of potential causes of failures in sensors

It's helpful to have a list of potential failures in sensor systems for reference in each application. Such a list also helps right from the start of hazard studies since there we are looking for ways in which the EUC control system can fail and create a hazardous event.

This list is compiled from experience and from a useful listing available in Appendix B10 of ISA S84.01.

- **Components of the instrument**: Primary sensing elements are prone to failures due to extreme conditions. E.g. in pressure transmitters overpressure stresses caused by pulsation or by hydraulic testing without isolation. Diaphragm seals with filled systems may slowly leak. Drifting occurs if excessive vacuum conditions arise. See manufacturer's limiting conditions for diaphragms.
- **Process connection faults**: Impulse line leakage or blockage or condensation builds up. Wrong location of pressure sensors – incorrect pressure conditions. Wrong location of temperature sensors – incorrect representation of actual temperature. Thermocouple incorrectly seated in thermowell. False readings caused by changed process conditions: examples:
  Density changes in tank fluids lead to errors in hydrostatic head level measurements; Aeration of fluid, foaming, temperature changes
  Flow meter errors due to entrained air or gases, partly filled pipelines
- **Accidental isolation**: process connection isolator valve left shut after maintenance.
- **Hostile process conditions**: Fouling/corrosion/viscosity/clogging/pulsation. Malfunction of proximity switches and other solid-state switch devices may arise due to off-state leakage rates being too high. The leakage current may be higher than the off state threshold of the logic solver input stage. Inductive or capacitive coupling may falsely turn on inputs to a logic solver.
- **Wiring errors**: Wrong sensor pair used after maintenance, loose terminals, corrosion of terminals, wire breakages due to stress, tangling, vibration.
- **Environmental mechanical**: overheating, freezing, mechanical distortion.
- **Environmental electrical**: RFI/EMI effects on instrument or on long transmission lines inducing offsets. X-ray interference on nucleonic detectors.
- **Specification/range/resolution**: Wrong specification. Smart transmitters incorrectly adjusted by remote configuration. Accidentally left with forced output condition.
- **Calibration errors**: Maintenance error, misreading of data sheet, faulty calibrator or instrument left on test mode.
- **Response time:** Too slow for process conditions; too fast causing spurious trips on short surges.
- **Power supplies**: Failures or out of spec. Accidental grounding leading to fuse failures.
- **Intrinsic safety barriers** (additional devices in the loop).

- **Lightning damage**. May destroy sensors or may lead to increased leakage in electronic circuits, hence major calibration errors.

7.4.3   Failure modes

Summarizing the effects of the potential causes of failures we can see that there will be a diversity of failure modes for sensors. We have to bear in mind that there are only two classes of failure as far as the safety system is concerned; i.e. dangerous undetected failures or safe detected failures. Both types are undesirable but our first target is to minimize the dangerous undetected failures.

Type	Signal	Failures/Comments
Analog sensor/transmitter	4–20 mA or 0–10 V dc	Upscale, downscale, freeze or drift. Zero shift or sensitivity change
Intelligent Transmitter	4–20 mA or	As above but also unpredictable due to programmable electronics.
Intelligent or Smart Transmitter	Field bus	Digital transmission and multidrop configurations may cause unpredictable failure modes unless special protection techniques are added.
Thermocouple	0–100 mV	Open circuit (burnout). Converters configurable to upscale or downscale on burnout.
RTD	100 ohms	Open circuit, upscale Short circuit, downscale. Converters configurable to upscale or downscale on burnout. High resistance terminals may cause calibration errors.
Current/voltage alarm trips	Converting 4–20 mA to discrete 0 or 1 logicals for input to SIS	High or low, unpredictable. Field adjustable trip values are not secure. Fail-safe on power supply failures.
Panel based trip amplifiers	As above. Responses to sensor failure modes are configurable	High or low. Also affected by power supply failures. Beware of RFI problems causing spurious trips.
Switch sensors	0–24 volts Dc or Ac	Stuck on condition. Fail to danger mode not detected until hazard demand occurs. 1oo2 redundancy desirable.

Table 7.2
Some typical failure modes of instruments

Before we look at ways of overcoming the potential failures let's take a similar look at actuators.

7.4.4   Actuator types

The term actuator can be misleading. The devices used by the SIS to actuate the protective function should more correctly be called final elements.

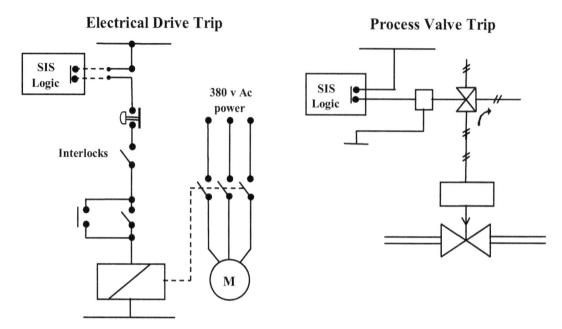

Figure 7.4
Two common types of final control elements

Final elements are most commonly seen as process valves with air or hydraulic power to actuators that are spring loaded to close on release of the actuator fluid. (Typically described as 'air to open, spring to close'.) Or they are simply motor starter contactors that must be de-energized to break the power supply to the drives that must be tripped.

In large power control applications or in large valve applications the final elements may have to use active power to carry out its trip task. These applications require backed up (i.e. redundant) power supplies to operate heavy duty power isolating contactors or to drive motorized valves. In such cases the power system becomes an integral part of the final element and would require diagnostic monitoring.

In all cases there must obviously be a high degree of assurance that the final element will do the job when called upon to act. Hence there will be an emphasis on diagnostics and regular proof testing.

Remotely operated shut-off valves
In chemical and oil processing plants there is often a need to protect the plant and environment from the release of a hazardous substance or to stop the incoming feed such as a gas supply. Fail-safe isolation valves are required to respond quickly and be able to provide effective leak tightness throughout any emergency. This usually means the valves must be firesafe; a form of construction and sealing that prevents pipeline fires from breaking through the valve.

These valves are often designed to be operated remotely either by operator action at a panel in a safe location or by an automatic trip system. Similar valves are used on offshore platforms where they are known as ESDVs. Such valves are effectively final elements in a functional safety system and all fall within the same guidelines applicable to safety instrumented systems.

A useful introduction to the subject of emergency isolation of process plants is obtainable from the UK Health and Safety Executive via their website: www.open.gov.uk/hse/hsehome.htm.

Potential failures in final control elements

The table lists some typical threats to the correct operation of a safety shutdown valve. It is important to bear in mind that the valve operation depends on several components to work properly. These are typically:

- The output switching signal from the logic solver
- The solenoid valve used to feed and vent the power air or hydraulics
- The feed and vent tubing to the actuator
- The actuator itself
- The valve stem and seat

These items must be carefully checked in the design and in the continuing maintenance of the valve. Some possible failure modes are shown next.

Type	Signal	Failures/Comments	
Process isolating or relieving valves	Air or Hydraulic power. Routed via solenoid valves to the actuator. On/off switching is normal. Some applications may use position control signalling to ramp valves to a safe setting.	1	Open, closed or partially open (frozen or stalled).
		2	Leakages due to erosion/corrosion of seats or failures of seals.
		3	Insufficient actuator force to close against pressure or flow.
		4	Actuator air lines can become trapped and blocked stopping the venting action.
Solenoid valves	Power control signal from SIS output relay.	5	Stuck open after long periods
		6	Pilot driven valves prone to blockages, stick open.
		7	Direct acting solenoids with spring action to block supply and to vent actuator are preferred.
Electrical drive controls	MCC contactors supplying power to drives, heaters or VSDs have their control power cut by a trip relay contact from the SIS.	Contactors have a very low fail to danger rate and are not normally required to have any redundancy.	
Output trip relays	Power to coil, on/off.	Most likely to fail in de-energized condition. (fail safe) Coil inoperative, contacts stuck open or closed, contacts welded closed by arcing, high resistance.	
Solid-state relays	On/off logic.	Not recommended due to potential failure mode on.	

Table 7.3
Summary of failure modes for final control element

7.5   Guidelines for the application of field devices

The application of any measurement and control device for duties in a safety instrumented system must take into account two primary considerations:

- The device must be applied using the best design techniques to minimize failures.
- The device should meet the design requirements of IEC 61508 or IEC 61511.

7.5.1   Design techniques to minimize failures

Given the diversity of failure modes the design techniques required to minimize the fail-to-danger rate will be application dependent.

The ground rules for design to minimize dangerous failure rates include the following techniques:

- Fail-safe design
  Design sensors and actuators to result in fail-safe responses to their most likely failure modes. Then review spurious trip rates to see if they are acceptable.
- Separation
  Ensure separation between BPCS and SIS sensor/actuator systems as far as practicable.
- Diagnostics
  Search for ways of introducing diagnostics to frequently confirm the healthy operation of the device.
- Redundancy
  Use redundancy where a reduction in fail-to-danger rate is needed or where a low spurious trip rate is essential.
- Diversity
  Search for diversity of sensors where the risk of common cause failures is significant.

Let's take a brief look at each of the features listed above:

7.5.2   Design for fail-safe operation

- Sensor contacts closed during normal operation
- Tx signals go to trip state upon failure (normally < 4 mA)
- Broken wire = trip
- Output contacts closed and energized for normal operation
- Final trip valves go to trip (safe) position on air failure
- Drives go to stop on trip or SIS signal failure

The points summarized in the above list are fundamental to good SIS design practices. There are however no set rules as each installation must be evaluated to see if the majority of possible failure modes will lead to a safe condition for the process. For example, the standards allow us to use energize-to-trip designs if needed, but require that these achieve a high safe failure fraction through the use of diagnostic techniques such as pilot current sensing.

7.5.3 Separation of sensors from BPCS

One of the ground rules of safety instrumented systems is:

Avoid sharing sensors with the BPCS (basic process control system) except under specially reviewed conditions.

The reasons for this should be fairly obvious but it is surprising how often the rule is broken due to financial or physical constraints.

Do not share sensors because it:

- Violates the principles of independence for the SIS from the BPCS
- Creates potential for common cause failure
- Does not create a separate layer of protection
- Procedures for maintenance and testing and device protection may not be adequate.

The following two graphics illustrate the sharing and separation of sensors in a typical boiler drum level application.

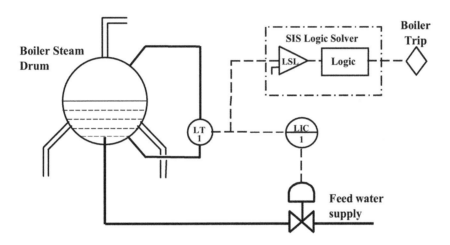

Shared sensor for control and trip: Not acceptable

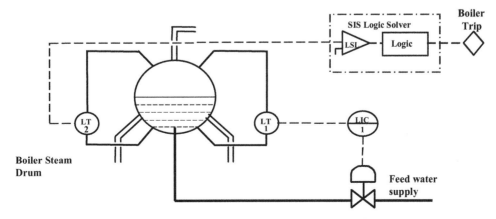

Separate sensor for control and trip: Acceptable

Figure 7.5
Sharing and separation of sensors

Note how separation includes separate connections into the process; a contribution to reducing common cause failure potential. The next diagram presents a fault tree analysis comparing the effects of shared and separate sensors on the damage rate.

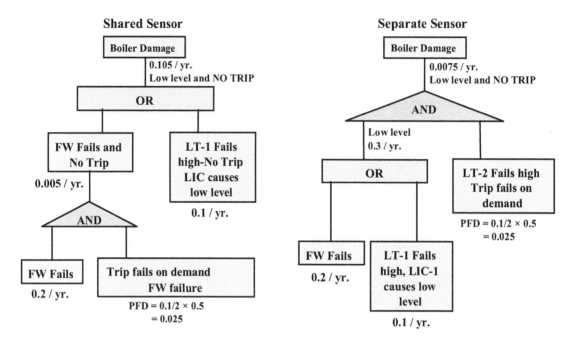

Figure 7.6
Fault tree analysis showing how separate sensors improve hazardous event rate

Note that the fault tree on the left separates the faults due to feed water failures from those due to the transmitter failure. No protection is provided against transmitter failure, which will cause a low level condition through its action on the level control loop LIC-1. Hence the boiler damage rate will be high compared with the correctly designed version on the right-hand side.

Limited exceptions to the separation rules
Separation rules: Sensors
Under the earlier ISA guidelines and in the new IEC 61511 standard, there are permitted exceptions to the separation rules for sensors and actuators. If you can show that the safety integrity requirements can still be met it is permissible to share sensors in an SIL 1 application. This would only be possible if the sensor was used in the BPCS for a control function that was not related to the task required in the SIS; i.e. if when the sensor failed it did not contribute to the hazard demand placed on the SIS. In practice this is unlikely to be true unless the sensor is working as a recording device.

Even in this instance the sensor should be installed primarily as part of the SIS with all its protection and test facilities, and then have its value copied across to the DCS via a loop isolator device. Additional diagnostic will also be needed to meet reliability targets.

For higher SILs there is no question that separation from the BPCS is essential.

Separation rules: final elements
The rules are slightly different for the sharing of final elements. In the case of electrical drive tripping it is normal practice in SIL 1 and SIL 2 applications to trip drives by

removing the power to the main control contactor (as shown in paragraph 7.4.4). For SIL3 the need for redundancy dictates that a second contactor should be used.

- A single valve may be used for both BPCS and SIS provided that:
  a) Valve failure does not create a demand on the SIS

  b) Diagnostic coverage and reaction time are sufficient to meet safety integrity requirments

- Recommendations for a single valve application as shown in the next diagram
- SIL 2 and SIL 3 normally require diverse separation as shown in the subsequent diagram

For shutdown valves it is acceptable for SIL 1 systems to use an existing process control valve to trip the process but only if the hazard analysis shows that faults in this valve are unlikely to place a demand on the SIS function.

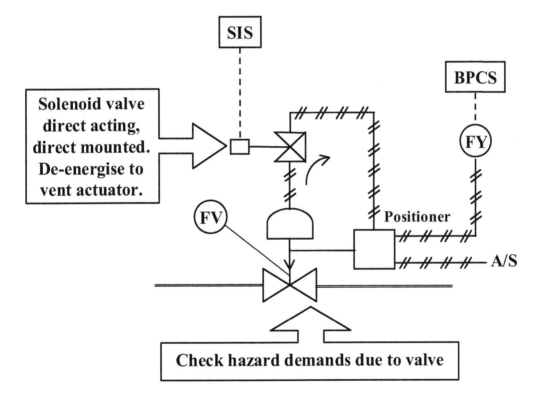

Figure 7.7
Arrangement for tripping of shared control valve in a SIL 1 design

IEC 61511 specifically calls for the trip solenoid to be arranged as shown in the next diagram. The arrangement shown in the diagram ensures that positioner defects do not prevent the trip action from being executed.

For SIL 2 applications the preferred method to achieve separation is to install a separate shutdown valve specifically for shut-off or venting duties. This duty is often met by a ball or gate valve.

The redundancy requirements of SIL 2 and even SIL 3 can often be met by also tripping the control valve, leading to the arrangement shown in the next diagram.

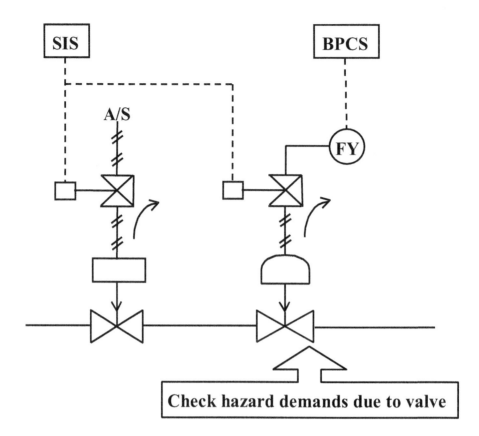

Figure 7.8
Diverse separation of control and shutdown valves suitable for SIL 2 and SIL 3

In most cases the decision on how to arrange the final elements will have a major cost and performance implication. This is an area where careful application of fault tree analysis can be very helpful.

7.5.4   Sensor diagnostics

Diagnostics as we have seen before involve the monitoring at frequent intervals of the operating condition of a device such as a PES or a sensor. Thus a high level of diagnostic coverage will convert many potentially dangerous failure conditions into a safe condition with an alarm or safe shutdown.

Diagnostics can be applied as self-testing techniques for sensors and valves in addition to functional testing or proof testing which is carried out periodically as part of the overall maintenance of the SIS.

Before the days of the microprocessor diagnostics were not normally applied at all to sensors and actuators. Programmable systems allow us to organize improved methods of self testing the sensors and actuators. Here are some of the possibilities.

- Do not confuse with proof testing
- Compare trip transmitter value with related variables; not often practicable
- Use safety certified transmitters, if available
- Use Smart transmitters with diagnostic alarms
- Use redundant transmitters to compare values.

Proof testing cannot be claimed as a diagnostic due to the very low frequency of testing. Smart and safety certified transmitters provide diagnostic monitoring of their own electronics and they can report on faults or out of range signals. Hence the measurement system becomes to some extent fault tolerant. The question is what level of diagnostic coverage is achieved? There is still scope for measurement error due to impulse line problems or changed process conditions. More on this later.

7.5.5   Valve diagnostics

We need assurance that a valve that may have been held open by air or hydraulic pressure for a year or more will close at the right speed as soon as the trip condition is detected. On-line testing is required but in most cases we cannot just trip the valve closed. Diagnostics seek to prove or infer that the valve is working properly. This is not simply a matter of seeing the valve move because there may be symptoms of trouble to come even though the valve is able to move. We need to check for freedom of movement, proper venting of the actuator, normal response time and absence of any tendency to stick or hang up. Here are some established methods; the choice is naturally dependent on the application:

- On-line trip testing: not really a diagnostic but a proof test. However if the valve can be arranged to trip fully shut (or open) without shutting down the process, perhaps by using parallel paths or bypasses then this test can be performed at regular intervals to improve the PFD of the valve.
- Discrepancy alarm. Uses a limit switch feedback signal; the logic solver generates an alarm if the valve position does not match the command.
- Trip testing with position feedback: A position signal from the valve will allow the tester to observe the test response rate and hence verify correct movement characteristics. If a PES logic solver is used it can be arranged to record or even check the response profile and detect deviations from the standard version.
- Partial closure testing: The same as trip testing except that the valve trip is performed automatically by the logic solver and then cancelled before the valve can move more than say 25%. The position feedback can be as simple as a limit switch, but a position transmitter is better; the disturbance to the process is minimal.
- Use a smart valve positioner to do all of the above tests and more. A smart device allows the partial closure test to be done automatically by the actuator positioner control system independently of the logic solver. It also allows the actuator pressure/time responses to be 'fingerprinted' and checked for major variations between tests. More on this in the 'Technology' section.

Offshore installation practices for ESD valves

One particular example of valve proof testing and diagnostic requirements is found in the mandatory provision of ESD valves on offshore oil and gas platforms. These are required to ensure safe isolation of the platform from incoming flows of oil or gas. They are critical to the safety of personnel on the platform since a fire on board could be fed by a high pressure feed of fuel from the undersea pipelines or wells.

Safety critical valves are subject to extensive inspection and maintenance requirements. Every pipeline riser connected to the platform must have an EDSV with a local closure control panel that ensures the ESDV remains closed once it has been tripped by an operator or by the platform ESDV system (i.e. the SIS). Annual trip testing is compulsory for such valves as well as a partial closure test every alternate six months. In this case the closure tests are performed from a local control panel with visual inspection of the response. Clearly an automated version of this activity is desirable.

7.5.6 Redundancy in sensors and actuators

It is often quite difficult to arrive at a decision on redundancy in field device applications. There are conflicting demands between safety, reliability, cost and regulatory requirements. Let's look at the constraints and find some guidelines.

- IEC 61508 part 2 places an upper limit on the SIL that can be claimed for any safety function on the basis of the fault tolerance of the sub systems that it uses.
- Limit is a function of the hw fault tolerance.
- The safe failure fraction.
- The degree of confidence in the behavior under fault conditions.

Constraints applied by IEC standards: IEC 61511 sets down minimum hardware fault tolerances where the safe failure fraction of an instrument is between 60% and 90%. Fault Tolerance (*FT*) is the ability of the system or sub system to provide protection in the presence of a fault; hence redundancy increases fault tolerance. An outline of the requirements is shown in the next diagram.

- IEC defines two types of equipment for use in safety systems:
- *Type A: Simple Devices: Non PES*
- *Type B: Complex Devices: Including PES*

The acceptable fault tolerance rating of *B* is less than *A* except under certain conditions as can be seen in Table 7.4 below.

It is important to note the distinction between simple (non-PES) and complex (PES) devices. This places a penalty on instruments where there is less confidence about the failure modes due to the use of programmable electronics (PES).

The table indicates a greater requirement for redundancy where PES devices such as smart transmitters are used but exceptions apply that effectively allow us to use most types of 'smart' transmitters just as if they were conventional analog transmitters.

Exceptions:

- Reduce fault tolerance (FT) by 1 if SFF > 90% (e.g. use diagnostics to achieve this); OR
- Reduce FT by 1 if the instrument is 'proven in use' AND has a limited configuration facility AND is password protected AND SIL is below 4.
- Increase FT by 1 if SFF < 60% (normally applies to a non-fail-safe device; to be avoided anyway).

For the meaning of '*proven in use*' please refer to the next sub-section on selection of instruments.

Hence when considering the choice of sensor, bear in mind that if it does not meet the best credentials it may be necessary to install redundant units to meet the SIL targets.

Fault Tolerance of Sub-Systems.
IEC Requirements: From IEC 61511 – part 1. Clause 11.4.
Applies to Sensors and Actuators with Safe Failure Fractions 60% - 90%.

Safety Integrity.	Simple Devices (Non PES)		Complex Devices (Using PES / Smart etc.)	
	Min. Fault tolerance.	Min. Architecture	Min. Fault tolerance.	Min. Architecture
SIL 1	0	1oo1	0	1oo1
SIL 2	0	1oo1	1	1oo2 or 2oo3
SIL 3	1	1oo2 or 2oo3	2	1oo3
SIL 4	2	1oo3	Special requirements apply, see IEC 61508	

Exceptions apply for higher and lower safe failure fractions – see text.

Table 7.4
Fault tolerance of sub systems

Constraints applied by SIL target failure rates: If the device on its own has a poor PFD figure (i.e. relatively high compared to the SIL target value) it may be necessary to use a redundant 1oo2 or 1oo3 configuration to meet the target.

Constraints applied by spurious trip target rates: The spurious trip rate of a 1oo2 pair can often be 10 times the dangerous failure rate. If this is too costly for production losses the logical thing to do is install a 2oo3 configuration. These configurations are very commonly seen in process plant applications, but we must always keep in mind the extent of common cause failure potential or systematic design errors such as wrong calibration of all three transmitters. Reduction of common cause requires more diversity as described in the next section.

Redundancy summary: See the next two diagrams for an outline of the basic redundancy options.

Sensor or Actuator Config.	Selection
1oo1	Use if both PFD and nuisance trip targets are met.
1oo2	2 Sensors installed, 1 required to trip. PFD value improved, nuisance trip rate doubled.
2oo3	3 Sensors installed, 2 required to trip. PFD improved over 1oo1, nuisance trip rate dramatically reduced.

Table 7.5
Summary of the most commonly used redundancy options

Table 7.5 sets out the basic rules for choosing between single channel or the most commonly used redundant channel operations: 1oo2 and 2oo3.

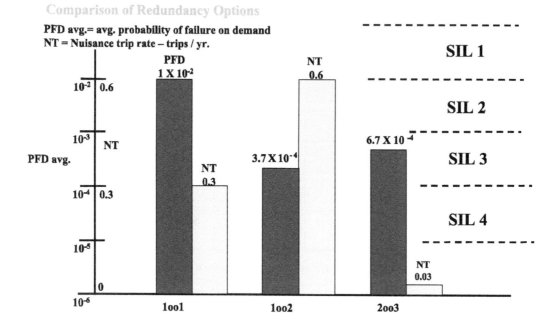

Figure 7.9
Comparison of redundancy options

Figure 7.9 gives an indication of the impact of redundancy options for a typical field device application. The data given are based on a sensor system with failure rate of 0.05/yr and a safe failure fraction of 75%. Note how the spurious trip is drastically improved by using 2oo3.

Sensor or Actuator Config.	Selection
1oo1D	Internal and external diagnostics used to improve safe failure fraction. Alternative to 1oo2 for SIL2
1oo2D	As for 1oo1D but able to tolerate 1 fault and revert to 1oo1D during repair. Meets SIL3 if safe failure fraction exceeds 90%. Does not satisfy diversity for SIL3 if sensors are identical. Reduces spurious trip rate, good alternative to 2oo3

Table 7.6
Options for minimizing field device redundancy through diagnostics

Table 7.6 shows that potential exists to reduce the amount of redundancy needed in field devices by using automatic diagnostics. If the safe failure fraction can be increased above 90% by using internal and external (i.e. in the logic solver) diagnostics the redundancy rules allow one less level of redundancy. In particular the 1oo2D configuration used with sensors means that spurious trips can be avoided by allowing the system to degrade to 1oo1D when a defect is reported in one of the sensors. The defective sensor is then repaired whilst the plant continues to operate under increased surveillance.

Other points concerning redundancy
In the author's experience redundancy in instrumentation is often motivated by the knowledge that the basic sensor is in a difficult application. Hence the problem is not so much the reliability of the instrument as much as its technical performance in sensing. For example gas detectors looking for oxygen in flammable gas feed lines may have very drift-prone sensors due to contamination. In such cases a 2oo3 configuration is often used.

Another different scenario is seen in gas detector applications where a small leak is not considered serious until at least two locations respond with high levels of concentration. In this case voting is not for sensor reliability but rather for the logic of the physical conditions.

One advantage of using PES based logic solver systems is that they greatly increase the available decision making power to deal with redundancy options when compared with relay based systems.

Common cause failures
Common cause failures will limit the gains in reliability achieved through redundancy unless we take special measures to reduce them. There is always the danger that all our sensors or actuators will suffer the same defect at the same time. The sort of defects we have to watch out for are listed and illustrated below.

- Wrong specification
- Hardware or circuit design errors
- Environmental stress
- Shared process connections
- Wrong maintenance procedures
- Incorrect calibrators

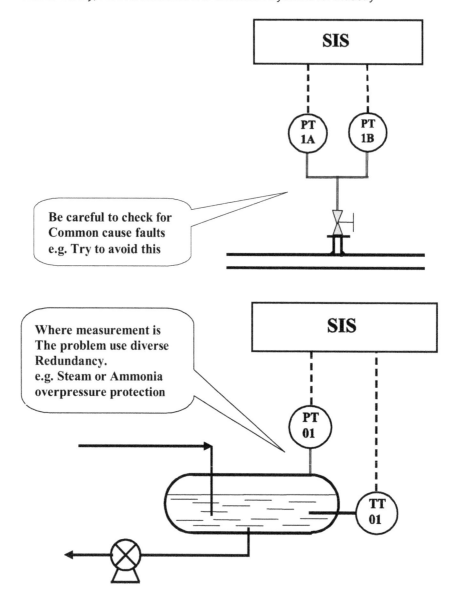

Figure 7.10
Common cause faults and the remedy of using diverse measurements

7.5.7 Diversity

Diversity means using different types of sensors and actuators or different measurement and operating parameters to achieve the same result in a safety system. The objective is to eliminate or minimize the possibilities of common cause failures or systematic errors. The previous diagram shows a typical example where the same limiting condition, in this case pressure in an ammonia storage vessel, can be detected by pressure and by temperature (which determines the saturation vapor pressure of the ammonia).

As SIL values rise so does the need for diversity in the protection systems. At SIL 3 the IEC standards expect us to provide redundant and diverse safety instrumented systems. At SIL 4 we should go to great lengths to avoid the risk of common cause failures by having redundant and diverse sensing systems connected to redundant and diverse logic solvers. One technique is to have a solid-state logic solver with a redundant PES logic solver.

7.6   Design requirements for field devices

Armed with our knowledge of failure possibilities and knowing the redundancy rules we should now be ready to specify and select particular types of instrument for any given application. The selection is almost entirely dependent on the application; hence it is not practicable to list any hard and fast rules about what instruments to use. Firstly we must remember that whatever instrument we use it must be qualified for use in our SIS.

7.6.1   Proven in use

We have already seen that the new standards want to ensure that we use instruments that have a good pedigree. They ask us to establish this either through design and quality assurance in accordance with IEC 61508 or because it is 'proven in use'.

What does 'proven in use' really mean? It is actually a specific term that was originally defined by the American Institute of Chemical Engineers. It has been incorporated into IEC 61511 part 1 as an alternative means of qualifying an instrument for SIS duties. Here is the definition found in IEC 61511:

A component may be considered as proven-in-use when a documented assessment has shown that there is appropriate evidence, based on the previous use of the component, that the component is suitable for use in a safety instrumented system.

IEC 61511-1 describes the requirements in some detail in 10 sub-clauses. Here is an attempt to summarize what it requires.

1) An assessment to be done to provide evidence that the component (e.g. instrument) is suitable for use in an SIS.

2) Assessment to include consideration of the manufacturer's QA system.

3) Assessment to include consideration of the performance of the device in a similar application.

4) Additional assessment items are needed where PES instruments (e.g. Smart) are used unless the configuration is limited via password protection allowing only process parameter settings to be changed.

5) Instrument must meet the functional and safety integrity requirements of the SRS.

6) Adequate documentation for integration, operation and maintenance to be available.

7) Assessment to be based on previous applications of the instrument, detailing similarities with the current application. Show that the likelihood of systematic faults is low.

8) Identify unused components of the instrument and show that they will not affect SIS.

9) Volume of operating experience to be valid for statistical confidence. Operational time of previously used components must exceed 1 year.

All of the above is intended to make sure that users do not use 'just any old instrument' in a safety system. In fact the problem lies with 'any new instrument' since it is difficult to establish valid previous operational experience. In particular we need to be aware that if we want to use an intelligent instrument in a safety application it has to meet the above requirements or it has to be certified that it has been built and tested in accordance with IEC 612508 or IEC 61511.

Problem 1

You wish to replace an old analog pressure transmitter with a new Smart equivalent, even from the same manufacturer. The Smart version does not have a 'safety certified' certificate from a testing house such as TUV. It will be necessary to find out from the vendor whether or not the instrument meets IEC 61508 or if it has been assessed by another party as being 'proven-in-use'.

Problem 2

You wish to upgrade your existing safety system to meet IEC 61508. Your existing sensors have done a good job and you wish to keep them. They are not PES based. You will carry out your own assessment report for 'proven-in-use'. It will be a lot easier if the instrument has a well-documented track record in your application both inside and outside your company.

7.6.2   Instrument selection

The IEC standards do not attempt to tell us what instruments to use. The ISA standard does offer some basic guidelines. The following points have been gleaned from the ISA standard and from personal experience.

Flow meters:

- Vortex shedding and magnetic flow meters preferred due to proven performance
- Head-type flow measurements to be avoided due to impulse line problems – leakage, condensate, freezing and drift

Temperature sensors:

- RTD and thermocouple types: Require burnout detection and alarm
- Be careful to locate sensors properly
- Ensure probes are seated in thermo wells
- Avoid thermistor types
- Infrared types with diagnostics

Pressure:

- Gauge, differential pressure and absolute pressure types all very reliable
- Ensure range and trip setting are compatible
- Ensure impulse lines do not risk condensate build up or blocking
- Use remote diaphragm seals instead of long leg links, but: beware of drift; beware of vacuum effects on diaphragms

Level: wide range of devices – good possibilities for diversity

- Check process for effects of aeration, entrained solids, density shift
- Differential pressure types are reliable but prone to process effects
- Ultrasonic and radar types have smart electronics, hence:
  a) Check compliance with IEC rules for PES
  b) Exploit diagnostics
  c) Check process for foaming, vapor, boiling
- Nucleonic types – reliable, often with diagnostics

Analyzers: reliability and proven performance

- Gas detectors often used in redundant modes
- Lower for diagnostic facilities
- Use comparative process signals where possible to support diagnostics

Limit switches:

- Use best quality industrial grade
- Sealed contacts where possible
- Proximity switches available with diagnostics
- Check leakage currents in OFF state to avoid false ON condition at SIS

Selection factors:

- Material compatibility
- Shut off duty, shock loadings, leakage, fire resistance
- Speed of response
- Element types most used are: Ball, gate, and globe
- Butterfly valves used on air systems, large sizes
- Spring / actuator performance margins. Fail close / open
- Requirements for diagnostics
- Requirements for limit switch and / or position transmitter
- Sharing of valve duties with BPCS
- Cost

Solenoids: critical components

- High grade versions only, stainless steel bodies
- Rated for outdoor service – sunshine to snow or enclosed
- Venting capacity to meet speed of response
- Use direct acting types – avoid pilot operated
- Use direct mounting, avoids risk of pinched vent lines.

7.6.3   Installation design features

- Direct connections to process for field sensors, separate taps and impulse lines.
- Dedicated cables, junction boxes, air lines and termination panels.
- Dedicated power supplies.
- Identification such as painting and labels.
- Devices to have local indication to assist proof testing.

It is important for keeping the safety integrity of field devices that details of the installation do not lead to accidental malfunctions. It is useless if a perfectly good pressure transmitter can be accidentally shorted out or cross connected to another circuit. Hence segregation of shutdown system wiring and special identification are features that will repay the extra effort involved. More points are covered in Chapter 8 as part of the general installation design guidelines.

7.7 Technology issues

This section considers the impact of new technologies on SIS field device practices. The two items that have a major impact on field devices are microprocessors and bus communications. We need to consider how these technologies fit in with SIS concepts.

7.7.1 Intelligent field devices: advantages and disadvantages

Intelligent instruments offer safety systems the advantages of being able to perform better quality measurements supported by internal diagnostics. We have seen how self-testing and reporting will help increase the safe failure fraction of a field device.

There are disadvantages for safety systems. Firstly there are the general reservations about the risks of programmable systems in safety applications. These relate to:

- Potential for systematic errors in the software
- User configurations may create new untested versions of the instrument
- Unauthorized in-service changes to settings, zero, range, mode etc.

One of the main purposes of IEC 61508 was to address these types of issues and find ways of dealing with them. Hence with the aid of safeguards based on IEC 61508 it becomes possible to use intelligent instruments in a safety system provided we stick to the rules.

In brief the answers to the above possible problems are:

1) Instruments using PES should be manufactured using hardware and software engineering procedures in accordance with IEC 61508.

2) A limited range of software instructions are made available to the end user to program the instrument within a tested range of configurations.

3) The program of the instruments is password protected.

We have seen that unless the above requirements are met, the fault tolerant rating of the instrument is loaded down by comparison with a non PES version. (See Redundancy section).

7.7.2 Application examples

Smart transmitters providing diagnostics:

Smart transmitters can be programmed to carry out internal diagnostics and then report any defects as soon as they occur into the SIS. The problem is to find a neat way to send the diagnostic status into the logic solver.

This diagram shows how a typical Hart transmission system will provide an analog input and a diagnostic input to the SIS. The Hart system is well established in the process industries and uses a frequency shift keyed digital signal superimposed on the standard 4–20 mA transmission. It can be used in a single point-to-point mode as shown in the diagram and if connected to a suitable gateway and decoder device a status contact can be used to provide an additional input to the logic solver indicating diagnostic status for the transmitter.

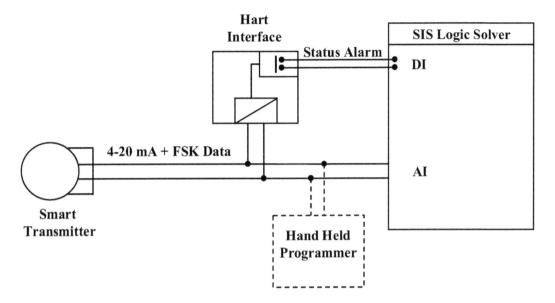

Figure 7.11
Hart transmission system used within an SIS application

Advantage: The systems use the diagnostic power of the Smart instrument

Disadvantage: The gateway device presents a new and possibly unproven failure potential in the safety loop. It is important to use a device that meets IEC 61508 requirements, preferably one that has been certified for safety use.

Another way to extract the diagnostic data from a Smart transmitter will be to use a full digital transmission system. The transmission system will have to be certified to have a very high safe failure fraction for it to be acceptable for SIS duties (as per IEC 61508 requirements). The logical extension of this principle will be to have a multidrop bus system dedicated to safety sensor duties as shown in the next diagram.

The arrangement indicated in this diagram allows a small group of sensors to serve a safety system whilst retaining the required safe failure characteristics. For a system such as this to qualify for safety duties it has to be manufactured and tested fully in accordance with the requirements of IEC 61508. It is basically a PES based functional safety system and is similar to the internal bus systems already certified for most safety PLCs.

The development of this type of technology has reached the stage where there are safety certified bus systems on the market. For example: the actuator sensor interface (ASI) system in the form supplied by Siemens is fully certified for safety duties in machinery applications. This system is relatively simple in that it is designed for binary devices only. See www.ad.siemens.de/safety.

For process control applications the Siemens S7-400F series of safety certified PLCs includes the Profisafe communication system which allows safety data to be transmitted within the Profibus DP message frames. Field networks are then permitted to transmit control and safety critical data over the same media. Remote I/O modules include an AS interface gateway certified for safety duties. The network software and safety interface products have been certified by TUV for SIL-3 duties in accordance with IEC 61508.

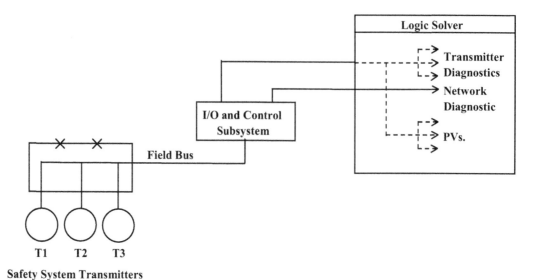

Figure 7.12
Safety system with field bus type communications via a fail-safe bus controller

Certification does not remove the need for the end user to decide if he can tolerate a failure of the control network and the safety system at the same time. This can be acceptable in machinery applications but requires great care in process situations. This area of technology is just emerging and it will take some time before its impact on process safety instrumentation practices will be clear. The position regarding the use of Foundation Fieldbus is also unclear at the moment as the original position adopted several years ago for ISA S84 was that Fieldbus should not be used in safety applications.

7.7.3   Safety critical transmitters and positioners

We take a brief look here at PES based field devices that have been certified in accordance with IEC 61508 for functional safety duties. Recall that we have a choice of proven-in-use or compliance with IEC. (We do not have this choice with PES logic solvers.)

What are the things that make a transmitter special for safety? The key features are:

- High diagnostic coverage and high safe failure fraction.
- Diverse redundancy principles applied internally.
- Proven software design and QA to IEC 61508 requirements.

For transmitters this will require the diagnostics to include ways of self-checking the measuring element within the instrument. If the self-checking can employ diverse and redundant measuring elements so much the better.

The diagram shows a schematic from the Siemens-Moore Safety Critical Transmitter: Note how a redundant and diverse measuring system is used to compare the internal values of the pressure sensor. Diagnostics driven by the PES section will shutdown this transmitter into a fail-safe state if a defect is found.

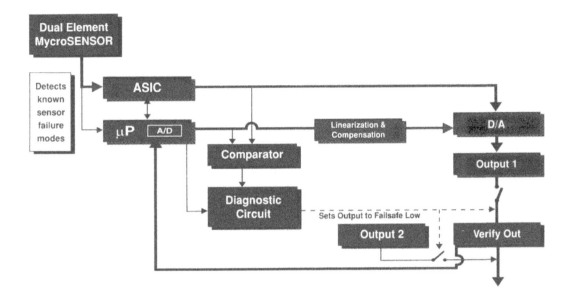

Figure 7.13
Diverse redundancy principles used in the Siemens-Moore Safety Critical Transmitter

The hardware and software engineering of the transmitter system has been certified by TUV to be in compliance with IEC 61508:

- Internal diagnostics with high coverage factor
- Very low PFDavg values. Saves on proof testing etc.
- Certified for single use in SIL 2 (instead of dual channel)
- Certified for dual redundant use in SIL 3 (instead of 2oo3)

The benefits shown in the diagram are those that we identified earlier in this chapter based on the high safe failure fraction and limited access by users to the software.

Safety certified positioner

The typical safety certified smart positioned would, automatically or on operator request, perform value response testing routines including partial closure tests to exercise a shutdown valve. Data returned to the DCS allows valve condition to be checked against reference values. Diagnostic tests within the positioner improve the coverage factor for dangerous failures.

This device operates as a control valve positioner but is also an intelligent device with Hart communications. As shown in the diagram the positioner copies its alarm and testing status into the Safety PLC or into a DCS.

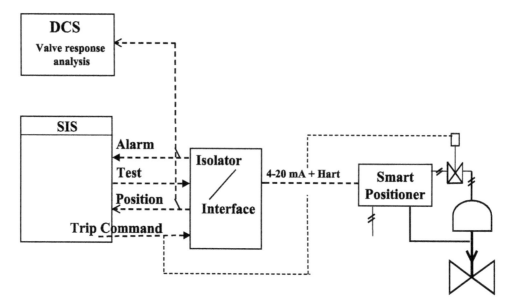

Figure 7.14
Control valve with safety certified smart positioner

The positioner can be programmed to carry out partial closure testing sequences at frequent intervals to verify that the valve mechanism is stroking correctly with the correct time/travel response. It can also measure the friction load in the actuator and with the aid of supporting software this can be used to predict maintenance needs.

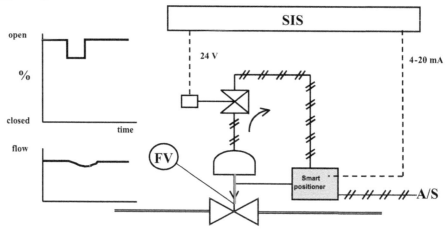

Figure 7.15
Automatic partial closure testing using a safety certified smart positioner or valve controller

As shown in Figure 7.15 partial closure tests allow the valve to be stroked away from its rest position without undue disturbance to the process.

The advantages claimed for this system include:

- Fewer valves to meet SIL 3
- Less piping and cabling
- Cost savings on manual proof testing
- Predictive maintenance benefits

The claim that smart positioner systems can reduce redundancy needs in shutdown valves is based on the assumption of increased safe failure fraction as described in paragraph 7.5.4. This conclusion would require thorough reliability analysis for each application before such a claim could be made, but the potential for major savings lies in this approach. What is certain is that frequent proof testing for typically 60% or 70% of possible dangerous failure modes of an actuator and valve combination provides a substantial improvement in PFDavg for the installed valve or valves. This may well be sufficient to deliver the SIL value required without additional measures and it will certainly reduce the required rate of manual proof testing with its attendant risk of production losses.

User experiences include particular benefits in applications where the process fluid encourages jamming of the valve.

Examples such as these show that new technologies in instrumentation are likely to make a substantial contribution to improved safety performance in field devices. The new safety standards provide a framework for dealing with PES based systems that will allow many more innovations to be introduced.

7.8 Summary of field devices for safety

This chapter has covered the key factors that influence the selection and arrangement of field devices for safety systems. Great care is required in the application of field devices because of their major contribution to the potential failure of a safety system.

The key points to remember are:

- Instruments must be well proven and have well defined failure modes.
- Intelligent instruments are treated as PES systems under IEC 61508.
- Follow well-established rules of separation, redundancy and diversity.
- Ensure instruments are properly qualified for safety applications through certification or proven-in-use.
- Strive to obtain good diagnostic coverage through logic solver power or self-testing schemes.
- Do not use bus technology unless a safety certified version is available.

8

Engineering the safety system: hardware

8.1 Introduction

This chapter looks at two aspects of engineering work for building an SIS. Firstly we look at some aspects of project engineering management and then we get back to some basic engineering practices. The objective is to discover some of the ground rules for building the SIS as a complete system.

As has been the practice earlier we shall check back with the safety life cycle model and the standards to see that we are doing the engineering work in the proper context. This raises the issue of what are the responsibilities of the project engineer and the user company? After a quick look at who does what we shall move on to a review of the detailed engineering practices mandated in the ISA standard. The points raised there constitute a useful run through many of the important ground rules of SIS design.

8.2 Project engineering

Project engineering and management play a major part in the delivery of a safety system. This section briefly considers the factors driving the need for good management of the project engineering activities.

8.2.1 Project problems

Both the UK HSE and Gruhn and Cheddie in their book have cited project management problems as contributing factors to failures in safety systems.

Causes of problems in projects are outlined. The points raised concern job timing and personnel.

The key points to work for here are:

- Keep the design team together and try to avoid changing personnel

- Communication between parties is essential and should be formalized
- Documented communications support the validation
- Set up a documentation plan with defined formats for each subject.

8.2.2 IEC requirements

Management of functional safety

It should be noted here that the IEC 61508 standard places much importance on the role of project management in the delivery of a safety instrumented system. In part 1 of the standard clause 6 provides two pages of items defining the management and technical activities that are '*necessary for the achievement of the required functional safety*'.

IEC 61508-1: Clause 6: Management of functional safety

Requires an organization to define the rules
  Subjects include:

- Management of functional safety in operating company
- Responsibilities of departments and individuals
- Documentation scope and systems
- Engineering procedures
- Management of software
- Competency and training of staff
- Management of changes

These items also specify the responsibilities of persons and organizations responsible for each phase of the safety life cycle. Altogether this clause of IEC presents the baseline material for a complete project management model for the safety project.

8.2.3 Functional safety assessment

The project engineering activities have to be formally defined and managed as part of the drive to achieve conformance with functional safety standards. Clause 8 of the IEC standard spells out the requirements for functional safety assessment by independent auditors as was outlined in Chapter 1.

8.2.4 Project engineering responsibilities

It is important that a project engineer is available to take overall responsibility for the engineering of the safety system. The PE should be able to bring to the job at least outline knowledge of most of the engineering essentials for the SIS. It is his or her responsibility to see that the job is done in compliance with the primary requirements.

- Management of the design phase and the installation phase is carried out within the framework of the SLC
- Appropriate and well-established engineering practices are deployed at all stages of the design
- Responsibilities are well defined

It is particularly important to define responsibilities in safety system project. The formal procedures of functional safety assessment require that each party in safety project has fulfilled its duties to ensure quality in the overall solution. Table 8.1 provides an example of a responsibility matrix for some aspects of safety project.

Participant	Functional safety management (FSM)	Define safety function (SRS)	SIS design	PFD/SIL calcs.	Validation
Operating company	A (See below)	Responsible	Approves	Provides data	Responsible
Engineering company	B	Contributes	Responsible	Responsible	Contributes
Instrument vendor	C		Contributes	Provides data	Provides certification
Logic solver vendor	D		Contributes	Provides data	Provides software validation

FSM Codes:

A: The operating company is responsible for all functional safety management (FSM) activities associated with the process, its operation and maintenance.

B: The engineering company is responsible for all FSM activities associated with the design procedures, the evaluation of the conceptual design, the detailed loop designs and specification of inspection, proof testing and maintenance requirements. The company is also responsible for FSM activities associated with the specification and implementation of application software requirements.

C: The field instrument vendor is responsible for the FSM activities associated with the instrument design, manufacturing, product care and the specification of inspection, proof and maintenance requirements. The vendor is also responsible for the provision of data supporting the claimed failure rates and proven in use records of the instruments.

D: The logic solver vendor is responsible for the FSM activities associated with the hardware and software design of the logic solver, the manufacturing and quality assurance procedures and for the provision of operating system software and compatible application programming tools.

Table 8.1
Typical responsibilities matrix for and SIS project

Note the roles of user, engineering contractor and instrument vendor with respect to defining the PFD values for process, loop and instruments. It helps a great deal in any project if one just makes sure that each of the parties listed is clear and committed about their duties.

The last time we looked at the IEC cycle we were at phase 5 dealing with allocating safety functions across the available protection layers and carrying out conceptual design of a suitable SIS. The reliability analysis helped us to decide if the proposed design was able to meet the required SIL and showed us ways of adjusting testing and configurations to meet the SIL.

Armed with an agreed (contractually approved) conceptual design and the safety requirements specification we should be able to go to the detailed engineering phase. Let's take a look at box 9 of the IEC standard.

The first thing we need to understand about box 9 of IEC 61508 is that it covers all the engineering activities associated with the specification, design, manufacture and testing of the SIS equipment itself except for the software engineering. It is a large and complex part of IEC 61508 but it is made a bit easier by having its own safety life cycle model. Hence the engineering activities are divided into:

- Safety Requirements Spec (we have already dealt with this in Chapter 4)
- Safety validation planning
- Design and development
- Integration and testing
- Installation, commissioning, operation and maintenance
- Safety validation

There is potential for confusion here because some of these activities are also applied to the overall safety life cycle for the plant (i.e. the combination of EUC, EUC control system and its safety systems). What we are looking at here is the actual safety instrumented system equipment.

8.3   Activities in box 9

The activities are arranged into a life cycle model as shown in Figure 8.1.

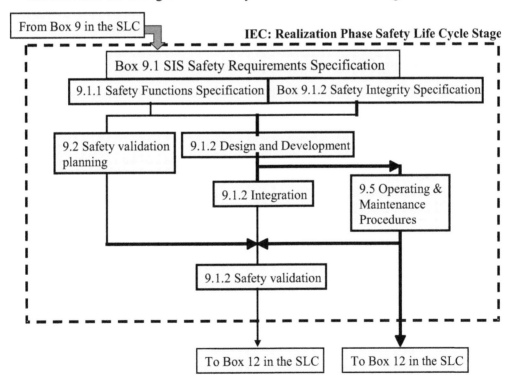

Figure 8.1
Life cycle model for the realization phase based on Figure 2 of IEC 615008 part 2

We have already dealt with the first items on the diagram (shown as Box 9.1) when we studied the safety requirements specification for the SIS and its component parts for functions (9.1.1) and integrity (9.1.2)

The next box is 9.2, covering safety validation planning, which tells us to plan the tests and records needed to demonstrate that the equipment will meet the SRS.

Box 9.3 covers the design and development of the E/E/PES system and is where most of the really significant design requirements of the standard are set down. This is the section where critical limitations on the choice of instruments, for example, are laid down. We have already looked at some of these points in the previous chapter concerning the architectural constraints for safety integrity (typically 1oo2 for SIL 2 and SIL 3) and the limitations on the choice of instrument subsystems (type *A* or type *B*). The details are generally beyond the scope of this book but here is the checklist of the critical design features covered in this part of the standard:

- Architectural constraints to meet the required hardware safety integrity
- Requirements to estimate the PFD due to random hardware failures
- Requirements for the avoidance of failures (relates to testing methods)
- Requirement for the control of systematic faults (relates to environment, human operator errors, software design errors, data communications errors)
- Evidence that the equipment is 'proven-in-use'
- Requirements for system behavior on detection of a fault

The details in the standard are applicable to all parts of the SIS and apply directly to persons building up a safety system from major parts such as instruments and logic solvers. When there is any doubt about what is the correct practice to use in a given design this section of IEC 61508 should be consulted.

Here are some key points arising from this part of IEC 61508.

8.3.1   Developing SIL for each application

The standard tells us that the method to be used is to decompose the SIS into major sub systems (e.g. sensors, logic solvers, actuators, also I/O sub systems) and apply the design rules for fault tolerance to ensure that each sub system meets the SIL value needed for the application. Then when we put the whole loop together we can work out if the overall SIS will meet the safety integrity requirements we have been asked to achieve. The SIL value we get will be limited to the lowest SIL value of the series connected sub systems as shown in the diagram below based on Figure 5 in IEC 61508 part 2.

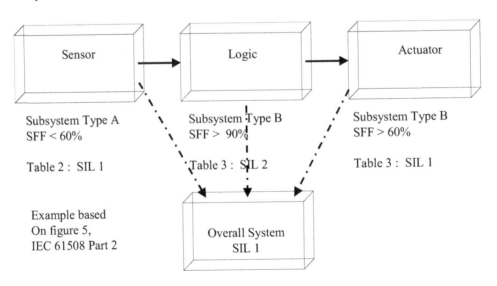

Figure 8.2
Overall SIL of a safety function depends on the lowest SIL of the components

The SIL value achievable by any sub system is determined by both its PFDavg for random hardware failures and by the level of fault tolerance required by the standard for a particular type of device. The standard defines these as architectural constraints in Tables 2 and 3 in part 2.

Architectural constraints

This part of IEC 61508 spells out the architectural constraints imposed on SIL values that we saw previously in Chapter 7. Recall that these require higher fault tolerance values in the sub system design (i.e. greater redundancy) unless a high safe failure fraction can be shown for the sub system. (See IEC 61508 part 2 clause 7.4.3.1).

Annex A2 of IEC 61508 part 2 contains tables indicating the degree of diagnostic coverage considered to be achievable in the various types of sub system. Various diagnostic methods are listed for each type of sub system (e.g. sensors) and each method is given a rating for the level of coverage likely to be achieved. This makes it possible for the designer to claim a certain level of coverage and hence define the hardware type for each sub system. From the data the designer can evaluate the required fault tolerance level of each sub system.

Safe failure fraction

In Annex C of IEC 61508 the standard explains how to calculate safe failure fraction for a sub system. This parameter is vitally important for the safety integrity or confidence level and as we have seen it will determine whether or not we have to build in greater levels of fault tolerance in a given application. The details lead to the basic relationship given here:

$$\text{Safe failure fraction} = \frac{\Sigma\lambda_S + \Sigma\lambda_{DD}}{\Sigma\lambda_S + \Sigma\lambda_D}$$

I.e. ratio of (all safe failures + all detected dangerous failures) to all possible failures. Recall that detected dangerous failures are found by diagnostics. Hence diagnostic coverage increases the value of λ_{DD} relative to λ_D. This shows how high diagnostic coverage can be used to increase the safe failure fraction of a sub system. This in turn may reduce the level of fault tolerance in the architecture needed to meet a SIL target. A good example of this is found in safety certified instruments where high coverage factors allow reduced levels of redundancy for SIL 2 and SIL 3 applications.

Integration

Box 9.4 covers integration or the 'bringing together' of all modules and sub systems making up the SIS and the testing of the finished product.

Operation and maintenance

Box 9.5 covers the operation and maintenance procedures required to keep the equipment working in its proper condition.

Safety validation

Box 9.6 covers the safety validation work and testing needed to prove the equipment will perform to meet the SRS.

Summary

In summary, Box 9 (or phase 9 as we describe it in this book) provides a comprehensive set of guidelines for:

- Design of the hardware sub systems to minimize random failures
- Selection of suitable devices
- Types of faults to be controlled by diagnostic measures
- Design activities to reduce the chances of systematic design errors.

Its limitation for the infrequent user is that its generalized terminology makes it difficult to use as a quick design guide. We need more sector standards to make things a bit easier to digest.

It may be helpful at this stage to take a look at the ISA S84 standard for guidance on detail design features. S84 has a more direct and practical set of measures to offer the instrumentation designer. These measures do not contradict the guidelines of IEC 61508 but are directed at typical process industry practices.

8.4   ISA clause 7: SIS detailed design

Mandatory requirements for design subjects

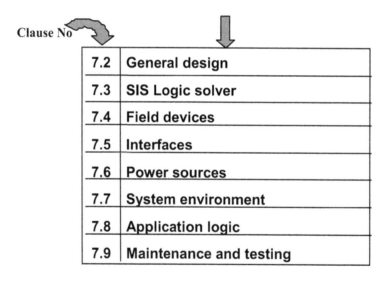

7.2	General design
7.3	SIS Logic solver
7.4	Field devices
7.5	Interfaces
7.6	Power sources
7.7	System environment
7.8	Application logic
7.9	Maintenance and testing

Table 8.2
SIS detailed design subject as arranged in ISA S84.01

The above table shows the topics covered in the detail design requirements that are described as mandatory. The intention in this chapter is to briefly run through some of the points found in the mandatory clauses. The items shown are in note form and are only approximate indications of the contents of some clauses. For the correct text and detailed interpretation it will of course be necessary to refer to the actual standard.

8.4.1   Clause 7.2 general requirements

The requirements include the following points:

The SIS

- Shall be capable of meeting the SIL
- May include sequencing functions
- May permit single or multiple safety functions
- Shall have documents and equipment under formal revision and release control
- Hardware and software to be compatible
- (Beware of revisions to firmware/software, also certification)
- Non-safety functions if implemented in SIS shall not interrupt or compromise any SIS safety functions (e.g. unclassified interlocks)
- Safe states of each SIS component to be defined
- Safety trips to stay tripped until reset. Reset requirements to be defined in the specification (SRS)

Note HW/SW compatible phrase. In practice this means that safety PLCs are only ever certified as an integrated HW/SW system and this extends to the programming package supplied with it.

Clause 7.2.10 reads as follows:

'7.2.10 manual means, independent of the logic solver, shall be provided to actuate the SIS final elements unless otherwise directed by the safety requirement specifications.'

Note particularly the requirement for manual means to trip outputs. This is optional by decision of the hazard review and SRS.

The next diagram shows a typical arrangement for satisfying the mandatory requirements for a manual means of trip in the offshore oil and gas industries. Note the complexity this can introduce since the SIS logic solver must know when a manually initiated trip has occurred.

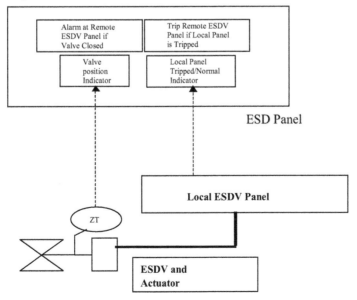

Figure 8.3
Local test and trip panel arrangement: offshore oil and gas industries

ISA clause 7.2 more general requirements

- Any single fault causing SIS failure, shall result in safe action
- Design to apply applicable codes/standards for environmental and hazard classifications; e.g. NFPA 70
- SIS input/output power circuits to be separated from all others (exceptions for shared sensors and elements)
- Provision for field bus if 'user approved'

Single fault clause applies typically as: 1oo1 to trip or 1oo2D to revert to 1oo1 on detection of fault in SIS.

8.4.2   ISA clause 7.3 logic solver

- Integrated design from supplier.
- Supplier to provide MTTF data, failure mode listing, covert failure rates. Method of data/source to be stated.
- PES to have diagnostics, embedded or application based.
- Should not allow process to re-start after power loss unless required.
- BPCS and logic solver to be separated. Exceptions for combination systems which meet SIL requirements (e.g. burner management, gas turbines).

8.4.3   ISA clause 7.4 field devices

ISA clause 7.4.1.1 (energize to trip vs de-energize to trip)

- Special cases for energize to trip designs.
  E.g. fire and gas protection systems, tripping of large electrical drives or power switchovers. Machine control systems.
- ISA calls for 'a method to assure circuit integrity' and suggests a pilot current sensing loop.

ISA Clause 7.4 field devices continued

Field Wiring

- Remote I/O systems: part of logic solver.
- Dedicated wiring loops for each field device.

The standard indicates that bus systems should not be used unless 'user approval' has been obtained.

Sensor Requirements

- Smart sensors to be write protected.
- Separation from BPCS sensors: 2 exceptions: redundant sensors, redundant protection.
- Sensor diagnostics to meet the SIL.

Final Element Requirements

- Separate shutdown valve for SIL 3.
- SIL 1 and 2 subject to review. Sharing with BPCS control valve is discouraged.
- Motors; sharing of starter with BPCS is acceptable.

8.4.4 Clause 7.5 interfaces

Human–machine interfaces
a) Operator
b) Maintenance/engineering
c) Communications

- Loss of interface shall not affect SIS operability. Provide backup means of operation.
- SIS status info to be available.

The general functions of interfaces are listed in this clause. An additional point of note is the general trend to integrate with operator interfaces in the BPCS. There is no objection to this practice provided the integrity of the SIS is not compromised. This point will be discussed in the book.

Additional points on interfaces given in ISA include:

- Operator interface not permitted to change SIS application software, use maintenance/engineering interface only.
- In batch related applications ISA permits BPCS or operator to select applicable logic function or recipe values. This requires security and selectable write functions.
- The scope of the maintenance interface:
  - Software manuals/tools
  - Programming terminals
  - Diagnostic tools
  - Test/bypass devices
  - Calibration devices
- Security of access to the maintenance interface must cover:
  - SIS operating modes, programs etc.
  - SIS diagnostics, fault handling
  - Add, delete or modify software
  - Data necessary for trouble shooting
  - Calibration devices
- Failure of interface shall not affect fail-safe operation of SIS.
  - Protect communications:
  - from other energy sources
  - provide dedicated power source

- provide shielding

- consider the use of fiber optics

8.4.5 Clause 7.6 power sources

Points listed in this clause include

- Design power sources to meet safety integrity and reliability
- Redundancy optional but strong case where 'energize-to-trip' is used
- UPS or battery backup options. Transfers must be bumpless
- Diagnostics to prevent use if SIS power is faulty
- E/PES power modules redundancy suggested
- Protect electronics against RFI/EMI
- Clean and surge protected power for PES
- Circuit breaker status copied to PES

I/O power distribution principles (B.7 not mandatory)

- No shared wiring loops
- Positive side protection against over current, over voltage
- Power isolation to individual devices for maintenance
- I/O power floating or grounded at one point. Eliminates stray paths to earth

A typical input circuit will be arranged to ensure that accidental grounding will force the input signal status to zero as shown here:

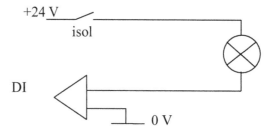

Figure 8.4
Input circuit configuration ensures safe failure if grounding occurs

8.4.6 Clause 7.7 system environment

- Protect against high/low temperature, humidity, vibration, RFI/EMI, hazardous area, flooding.
  NB. Avoid sprinkler routes in control rooms, plumbing and drainage runs, and leaking roofs. Common cause failure probabilities are high.
- Check environmental specs for SIS components.
- Improve conditions as needed to meet component specifications by using air conditioning and/or filtration.
- Panel ventilation and cooling via natural dissipation if possible.

8.4.7   Clause 7.8 application logic

Our next chapter looks at the requirements for application software in the logic solver. The ISA summary of requirements is given here for completeness.

ISA clause 7.8 application logic

- Electrical and electronic logic to use only logic controlled by the user under a formal revision and release control program.
- Clear and concise documentation system must be provided.
- All software in SIS must be under a formal revision and release control program.
- Manufacturer is responsible for formal revision and release control of embedded software.
- User is not permitted to modify SIS embedded or utility software.
- User must ensure application logic is documented in a clear and concise way.
- PES application software formal revision and release control program is to be maintained by the user.

8.4.8   Clause 7.9 maintenance or testing design requirements

This clause has some clear points to make about testing facilities:

Clause 7.9.1 requires the design to allow for testing of the overall system, which means being able to demonstrate the final element response to sensor operation corresponding to a trip condition. On-line testing facilities are required when the intervals between scheduled process downtime is greater than the proof test interval.

It is important to design in the testing facilities from the start and to do this we need to define the testing philosophy for each function.

Clause 7.9.2 calls for such facilities to be an integral part of the design. We want to avoid the maintenance team having makeshift arrangements for testing. The next diagram illustrates a permanent bypass arrangement for testing a shutdown valve on-line. Note the bypass warning 'ZA' alarm. This arrangement is used as a standard feature in some companies; others will avoid this as being unreliable or adding too much cost.

Clause 7.9.3 deals with test or bypass facilities and states that they should:

Conform to the testing requirements defined in the SRS.

As shown in Figure 8.5 always provide alarms or alerting procedures for the operator to know there is a bypass in force.

'Bypassing of any portion of the BPCS shall not result in loss of detection and/or annunciation of the conditions being monitored'. This last condition needs some care when the testing of a sensor is taking place.

Similar conditions apply to the possibility of forcing inputs and outputs (electrical simulation or software signal forcing). Clause 7.9.4 warns that this should not be done as part of an application or as part of an operating procedure or test procedure without special procedures supported by access security.

This completes the run through ISA design clauses. We should also recall that Annex B of ISA S84.01 provides a substantial amount of additional guidance that is not mandatory.

7.9.2 When on-line functional testing is required, test facilities shall be an integral part of the SIS design to test for covert failures

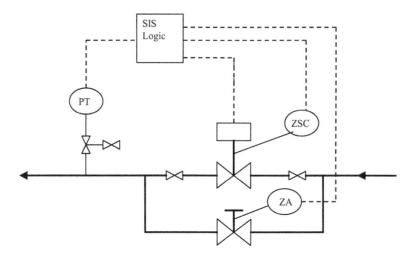

Figure 8.5
An arrangement for on-line testing of a shutdown valve using a bypass with alarm

8.5 Information flow and documents at the engineering stage

One aspect of engineering that is always needed is to plan the information flow required for the design stage of the SIS. Associated with this is the list of deliverable documents expected from the design team to define the manufacture and then to support maintenance.

This subject is best handled by a flow sheet of activities and sample documents.

For further guidance on the workflow needed for an SIS project there is a useful example set out in part 6 of IEC 61508. Annex A of part 6 is an 'informative' section on the application of parts 2 and 3 of the standard. Section A2 in particular describes the functional steps in the application of IEC 61508-2. Documents used or created in the realization phase will include:

Inputs

- Overall safety requirements specification.
- Safety allocations description with SIL requirements for each SIS function.

Develop

- SIS safety requirements specification.
- Safety validation plan.
- Hardware architecture diagrams with subsystems identified.
- Decision on logic solver type and vendor.
- Subsystem descriptions with parameters; e.g. proof test interval, diagnostic coverage factor, MTTR, safe failure fraction, PFDavg.
- Details of software requirements.
- Reliability model and SIL values achieved.
- Operating, maintenance and testing procedures.

Implementation

- Specifications of equipment for purchase with design requirements.
- Application software (see Chapter 9) documentation package.
- Integration test plans of HW/SW.
- Validation test procedures of each safety function in the logic solver.
- System interconnection diagrams and installation drawings, for final assembly at site.

8.6 Conclusion

The engineering rules for detail design are not particularly difficult or complicated except in the area of interpretation of the limitations on SILs. It is easy to develop the design details but the designer must be aware at all times of the context of the work, input from the SRS, output to satisfy the validation phase. The quality assurance environment is the key to success.

9

Engineering the application software

9.1 Introduction

The objective of this chapter is to provide some guidance on how to deal with the application software stages of an SIS project.

The chapter looks at some of the basic concepts and requirements that have been introduced in recent years to try and overcome the major concerns that have arisen over the use of software in safety applications.

The problem with software in safety systems is that it can be very difficult to control exactly what has gone into the system. The last thing we need in a safety system is an unpredictable response to a hazard demand.

We see how a systematic approach to the reduction of errors in specification and development has been laid out in the new IEC standard. The impact of this approach is found in the availability for end users of certified operating systems supported by certified programming packages. We look at some features of the packages and at the task of developing application software using them.

9.1.1 The problem with software

As the control system equipment moved from hardwired and solid-state equipment into PES safety practitioners found that the traditional approach to reliability analysis using quantitative methods to predict failure rates would not work with software. The initial response was 'if we can't measure it we mustn't use it'.

Fundamentally the problem with using software in safety system lies with the potential for systematic faults; i.e. faults that are not random such as component burnouts but have been introduced during the specification, design or development phases by errors in those processes. Such faults may then lie dormant until just the right combination of circumstances comes along.

One of the aspects of software that makes it particularly difficult to control is the ability for it to be re-used in new applications not originally intended by the designers. The tendency to use 'cut and paste' techniques to make up a new program creates a risk that the wrong features have been introduced to a new product.

It would help if it were possible to detect all systematic errors at the testing stages but it is well understood that there will be many combinations of logic and timing that cannot be fully explored in testing without a prohibitive cost in time and labor.

The situation was rescued by the development of software quality assurance. Management of quality in the production and testing of software replaced the quantitative assessment of reliability. The techniques originated with military practices and have become established in all areas where good software engineering is needed. This is not to say that all the problems have been solved but at least a broad agreement has been reached on a set of procedures that goes a long way towards creating a good level of confidence that the software in a safety system will perform to its intended purpose.

So, one of the main objectives of IEC 61508 is to lay down requirements and procedures for the specification, design, development and validation of software to be used in safety related applications. The methods used are consistent with the other parts of the standard; basically it sets out a software development life cycle model in parallel with the hardware development life cycle.

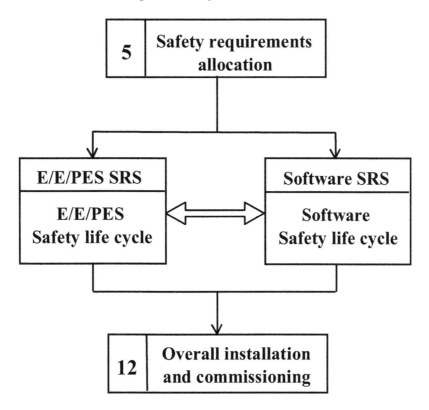

Figure 9.1
Safety requirements allocation divides into hardware and software components

The full scope of software safety life cycle activities is beyond the scope of this book. However, we need to understand what is required of the project engineer and the designers involved in practical applications. So here are some key points seen from the position of the end user.

9.1.2   End user position

From the practical point of view of the end user the position appears to be as follows:

1. Safety integrity requirements mean that the software installed in the completed SIS must have a proven quality assurance record.

2. The software in the SIS comprises two major parts:

- The embedded software or operating system (normally provided by the vendor of the PES)
- The application software for the specified safety functions (normally the responsibility of the engineering team for the SIS)

3. The validation of the finished SIS requires that a quality management system be correctly deployed and executed for both embedded and application software.

4. The easiest way for end users to achieve validation of embedded software is to purchase a fully certified PES from the vendor.

5. The best way to ensure quality in the application software is to follow the safety life cycle specification procedures and to use a certified application software tool, from the same vendor as the PES. New guideline material drafted for IEC-61511 part 2 provides greater clarity on the procedures to follow.

6. In cases where a specialized safety system is being developed for a particular product or packaged unit (for example: an elevator safety system) it may be necessary to carry out the software QA task over an entirely new PES design. In such cases the full software QA lifecycle approach may be needed as described in IEC 61508-3.

9.1.3   Basics of the software life cycle

We should now look more closely at the basics of the software life cycle activities. We begin by sorting out the main features of SIS software activities. For this we need to understand the architecture terms and relationships used in the IEC standards.

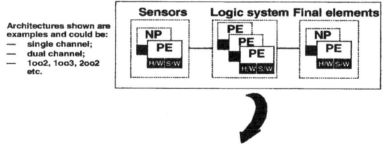

Figure 9.2
The software components of a programmable system

The above diagram, Figure 6 of IEC 61508, which describes how an SIS may have programmable and non-programmable parts. For each PES, for example the logic solver, there will be hardware and software components. The software components divide into the embedded part and the application part. Clearly the end user is interested mainly in the application software whilst the developer of a product such as a smart transmitter or a logic solver will be concerned with the embedded software.

Once the project team has established a safety requirement specification the life cycle activities needed to design and develop the SIS equipment will generally follow the life cycle activities shown in the next diagram.

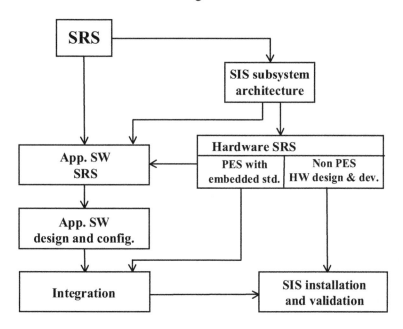

Figure 9.3
Realization phase activities for the SIS logic solver

Figure 9.3 shows the typical control system development routes in which hardware and software requirements flow from the SIS architecture decisions. We see that application software requirements are to be defined in the knowledge of the selected embedded software. In other words you cannot specify the details of your application without knowing the operating systems and utility tools you are going to work with.

At the end of the SW design and configuration work the application software will be integrated with the PES operating system and the actual hardware. Finally the completed SIS will be installed and linked up to any non-PES parts such as the actuators.

The software development activities are specified in IEC 61508 part 3 with a comprehensive range of measures and safeguards to reduce the chances of systematic errors. The framework for these activities is the well established '*V*' life cycle model shown in the next diagram.

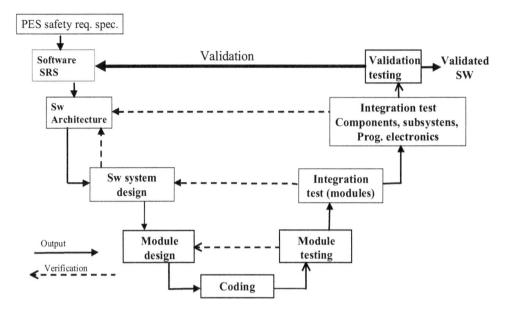

Figure 9.4
Principle of the 'V' life cycle mode for software development

The principle of the *V* model can be seen in the diagram where the descending levels of detail are matched by the ascending levels of testing and integration. There are some who are not convinced of the success of these concepts but they seem to be generally accepted as the best way to proceed.

9.1.4   Clause 7: Software safety life cycle

Clause 7 in IEC 61508 part 3 describes a full software safety life cycle model with 6 phases exactly as previously described for the hardware life cycle. Details of activities in each phase are specified and described in detail. These detail clauses provide very firm rules for the software developer and provide the basis of a quality assurance program for a new product.

9.1.5   Application software

To what extent do these QA techniques apply to application software?

This is a difficult question to answer. The IEC 61508 standard leaves it open to the user to interpret how much of the safety life cycle procedure should be used for application software.

There is a bit of a problem here for the typical process plant end user because this model and all the detailed requirements laid down in IEC 61508 part 3 are directed at situations where a completely new and customized product may have to be produced. Hence it is not always as clear how the end user should proceed in the case of a product with embedded software. However IEC 61511 helps to resolve this issue.

9.1.6   IEC 61511 provides guidance for end users

The upcoming process sector standard IEC 61511 is helpful in the case of end users because it shows how to proceed in the case where the embedded software is certified and provided by a product vendor. Part 1 of this standard has a clause where it specifies an application software safety life cycle. The model from the draft standard is shown in the next diagram (may be subject to revision before the standard is issued).

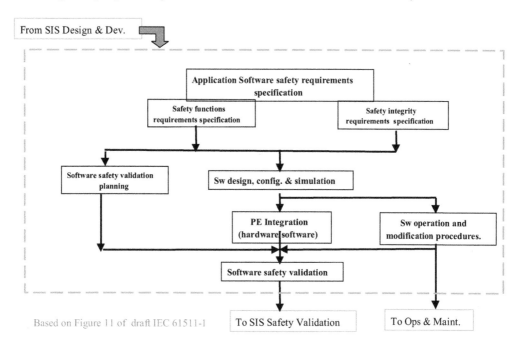

Figure 9.5
SIS design and development

This is familiar territory to any control engineer who has conducted a DCS or PLC application project.

9.1.7   Benefits of limited variability languages

Most helpfully for the end user is the concept of 'limited variability languages' or LVLs. These are programming language techniques that use high-level configurable program instructions such as ladder logic or function blocks to allow the end user to configure the application within a strictly controlled framework or environment. The standardization of techniques for LVLs and PLC programming languages was achieved through the application of IEC 61131, part 3 which defines the acceptable forms of language that are now in common use for PLCs.

Examples of function block configurations for simple input stages are shown in the next two diagrams. Typical arrangements for override facilities are re-used as proven stages and applied to many similar channels of signals. Once the end user has established standard procedures for dealing with most of the routine operations required in the safety system the benefits of re-using standard function blocks are quickly realized.

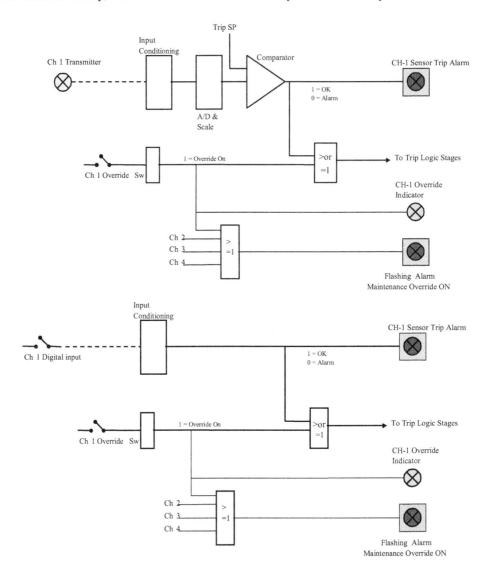

Figure 9.6
Function block configuration examples for analog sensor input (transmitter)

Essentially these languages fulfill the safety design requirement to use a minimal set of well proven and robust program modules in a standardized manner. Hence there is little or no opportunity for the end user to create a program condition that has not already been extensively tested.

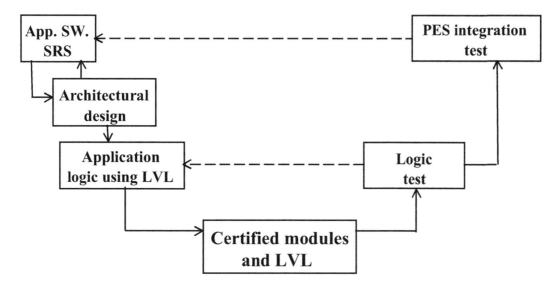

Figure 9.7
The 'V' life cycle model is simplified by using certified function modules and LVLs

Effectively the use of an LVL means that only the top two layers of the '*V*' life cycle model need be applied to the design, implementation, verification and validation of application software.

It looks as if this standard will be particularly helpful for project engineers conducting an SIS application project because it provides detailed requirements for all parts of the job.

IEC 61511: Application software life cycle requirements

- Application software safety requirements specification
- Features and facilities required of the application language
- Features to facilitate safe modification of the application
- Architecture of the application software
- Requirements for support tools, user manual and application languages
- Software development methods
- Software module testing
- Software integration testing
- Integration testing with the SIS sub system

9.1.8   Programming tools

For most process plant applications the engineer will require access to a configuration package supplied as part of the PES solution package. It may be helpful here to look at the basic features to look for in a programming package.

It is a fundamental requirement of a safety certified PLC that the integrity of the whole software package is assured. Thus the embedded software and the application tools should be developed by the same vendor and should be proven to meet all the constraints of the IEC standard when operated as a complete facility.

IEC 61511 part 2 describes the programming and support tools as 'the developer's workbench'.

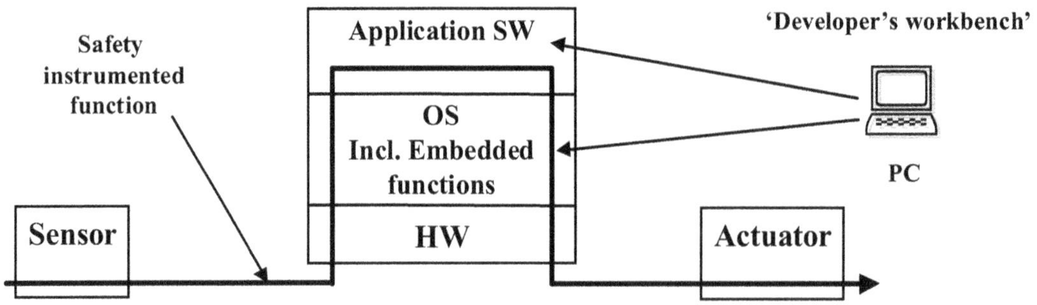

Figure 9.8
Information flow in the SIS and the role of the programming tool

Tasks

- Configures logic solver I/O and communication sub systems.
- Programs the logic and arithmetic functions of the application.
- Facilitates testing of applications.

Environment: Typically PC based running on PC operating system
The developer's workbench is the vital interface for the engineering phase of the SIS project and it assumes the role of maintenance support tool once the system has been installed. Commercially available logic solver packages typically include the facilities as described in the draft IEC 61511 part 2 and as listed below.

- Language editors for programming
- Function block libraries
- Special function block development tools
- Program execution tables and scan rate adjusters
- Downloading capability (to the logic solver)
- Emulation. Allows off-line testing of applications before downloading
- Program monitoring. Preferably in both the emulator and in the logic solver. On-line monitoring facilities may allow forcing of outputs or substitution of inputs under strict password control
- Diagnostic displays of the status of logic solver and all its modules

Some workbench systems also provide the ability to track program modifications between the version in the logic solver and the version in the workbench. This feature if properly validated reduces the testing work needed after a program modification.

The toolkit supplied by the vendor is often a critical selling point for the logic solver product since it can greatly affect the efficiency and hence labor costs of an application project.

9.2   Application software activity steps

Now that we have looked at the background to the software engineering task it may be helpful to run through a typical sequence of activities needed to deliver a fully tested logic solver application program from the initial specification.

9.2.1   Application software activities

This is an abbreviated version of the steps given in draft IEC 61511 part 2. It may be possible to link the data flow needed for these steps from the previous activities of the safety life cycle:

1. Configure I/O modules layout and memory variable data areas
2. Assign tag names to I/O and variables
3. Decide maintenance override philosophy
4. Define sensor and final element diagnostics methods. Also proof testing methods for I/O
5. Define external variables to be used for data feeds to DCS or supervisory systems
6. Develop any custom function blocks required to perform typical repetitive tasks. E.g. Input sensing logic with override
7. Ensure separation of safety critical programs from any non-SIS functions. Also ensure only certified function blocks are used for the SIS functions
8. Develop the application logic and functions ensuring correct order of execution of anytime or sequence dependent functions
9. Set up scan rates and sequence of execution
10. Use the workbench to emulate the operation of the programs
11. Download the accepted programs into the logic solver
12. Test all input/output functions and logic functionality using the workbench to monitor all signals. Also test the communication of data to/from the DCS or other systems.

The above sequence will of course be supported by documentation of the software and the testing procedures. Test procedures derive from the planning and validation activities specified for the safety life cycle activities. Documentation of the software will be supported by the PC based workbench as far as possible.

All software activities will fall within the scope of the project software quality management system, which will have to be set out to meet the requirements laid down in the IEC standard. Here is an outline:

9.2.2   Software quality management system

IEC 61508 part 3 clause 6 describes software configuration management. This is the task of housekeeping the project to meet the overall functional safety objectives. Subjects include:

- Controls to manage changes to software
- Maintenance of all documentation covering specifications, source code modules, test plans and results etc
- Change control procedures to process all modifications and track them through the design and testing phases
- Documenting the release of safety related software and the storage of master copies

9.2.3   Certification of operating systems

If the application software project is to achieve compliance with IEC 61508 it can only be carried out using embedded software and workbench software that has been certified for compliance with IEC 61508. Hence it is essential for the project to:

a) Use only hardware and software that is certified for safety critical applications.

b) Use software certified for the specific version of logic solver.

c) Ensure SIL certification for the hardware, software, communications interface and programming package is equal to or greater than the highest SIL required for the SIS application.

With regard to the certification of equipment we have already noted that authorized testing bodies such as TUV are used to certify specific vendor products. This applies equally to hardware and software. TUV certifies complete PES systems comprising hardware, diagnostics, embedded software, application software and programming systems. The certification is given for specifically defined products and is issued for defined integrity levels based on the VDE standard AK values and the IEC SIL values.

Scope of compliance for logic solver software products

The scope of compliance required for software products to be used in the application project is summarized here based on information in the draft IEC 61511 part 2.

- SIS logic solver and I/O certified for use at the relevant SIL.
- All of the programming languages supported by the logic solver with any special safety functions and function blocks to be certified for compliance at the relevant SIL.
- All restrictions and operating procedures required by the certifying organization to be stated in the user documentation.
- Methodology for on-line testing using overrides to be approved by the certifying organization.

9.2.4   Summary of software engineering

We have touched only the outline of what is a very large subject but the conclusions we can record at this level are as follows:

- Software safety integrity is achieved through life cycle quality assurance.
- Well-defined procedures exist for producing software for safety systems.
- IEC 61508 part 3 serves designers of new PES devices as well as end users. IEC 615 11 will make it clearer for end users.
- Certified software packages provide a secure platform for the end user to execute an application.
- Compliance with IEC standards requires end user commitment to safety life cycle procedures.

10

Overall planning: IEC phases 6, 7 and 8

10.1   Introduction

This chapter takes a brief look at the planning boxes marked in on the IEC safety life cycle. The idea of the planning stages being performed concurrently with the design appears to be a very good one.

It means that if you have a clear idea of how you are going to validate, install and operate the system at the time of starting the design you will be in a good position to provide the right facilities in the design 'the first time around'.

It helps us to foresee many of the requirements that we may not have thought of until it was too late.

In the next chapters we are going to look at some more details of installation, validation etc, so this is a good point to look at the scope of planning in each activity as suggested by the IEC standard.

Figure 10.1 shows the three planning boxes we have seen before and shows how they lead to their respective activities after the design and building of the SIS have been completed. It is important to be aware of the benefits of developing the plans concurrently with the detail design and building phase as listed in the next slide.

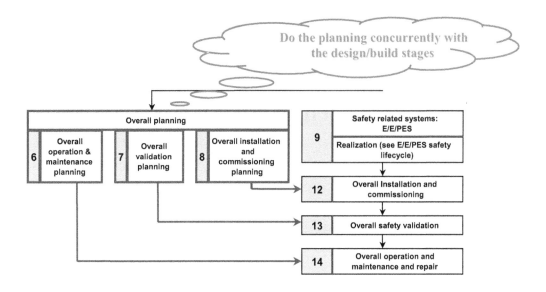

Figure 10.1
Planning functions in the IEC safety life cycle

10.1.1 Benefits of planning at the design stage

- Clarifies requirements for installation, validation, operation
- Helps foresee requirements for test facilities, overrides, controls, security
- Prepares the test and control documents whilst the knowledge is current
- Protects the SIS against project expediencies

This last point refers to the panic that may set in later in the job if time scales go astray. If the ground rules for the job are in place early on there should be no reason for project managers to try and shortcut the procedures for testing and validation.

10.2 Maintenance and operations planning

Maintainability and operability are key attributes of the SIS. It will be a poor solution if the regular testing and the upkeep of the system are in any way handicapped by shortcomings in the design.

10.2.1 What should we cover in maintenance and operations planning?

The short answer is everything you need to do to ensure the system is operable and that the safety integrity of the system is maintained. The specific (and longer) answer is set down in Box 6 of the IEC standard. Here follows a review of its main points.

10.2.2 IEC 61508 phase 6: overall operation and maintenance planning

A summary of the IEC requirements is as follows.
Phase 6 (Clause 7.7.2.1 in IEC 61508-1) has useful notes on items to be considered in the planning scope. See also the requirement for vendors to support software change

facilities. The following is a checklist for the planning scope based on notes provided in IEC phase 6:

- Routine actions to maintain the SIS. E.g. proof testing, calibration, cleaning, inspection
- Equipment/process operating procedures to avoid unsafe states and unnecessary demands on the SIS. (E.g. you should not cause the SIS to operate as a routine shutdown control for a plant. This is a feature often abused on plants)
- Constraints on EUC during a fault in the SIS (as discussed earlier)
- Constraints on EUC during maintenance of the SIS
- Conditions for removing constraints
- Procedures for restoring and confirming normal operation
- Circumstances for allowing bypasses or overrides to be used
- Permit to work procedures and authority levels
- Documentation for recording safety audits and proof tests
- Documents for recording hazardous incidents or 'near misses'
- Scope of maintenance work
- Actions in the case of a hazardous event
- Contents of an operating and maintenance log book

Design questions relating to operations and maintenance

This is the stage in a project when a set of questions should be set up to ask:

- How are we going to operate each safety function?
- Does it need start up procedures? Or will the safety function just sit there in the background? In particular:

'What do we do if the SIS is faulty and needs to be switched out whilst we do maintenance? Shall we shutdown the plant or can we go along with a safety clearance and put extra operators on the job to monitor the process whilst we run without a shutdown system?'

If the safety requirements specification has been properly done we would expect to know the answer to this question. It is worth noting that IEC 61511–1 has a draft statement to the effect that you may decide to continue running the process if you have alternative protection measures planned. It also says that if the repair period exceeds the specified MTTR you are required to take a specific course of action designed to ensure safety. Usually this means a manually initiated shutdown.

When those rules are decided there will be a set of work procedures that define the operating tasks and how they are to be carried out. What this standard is asking us to do is draw up those procedures now whilst the detail design is starting.

An example of the impact of operating procedures on the detail design of the SIS is given in the next paragraph.

Impact of operating procedures on SIS design

Operating procedures define:

- The content and presentation of user interfaces such as

- Format and content of video display screens
- Layout and content of indicators, alarms and switches on a panel based user interface
- User interface procedures for controls such as process overrides, reset functions, timers and set point changes
- Recording/printing details for safety related event logging and alarming functions
- Security controls and displays for setting and cancellation of maintenance overrides/bypasses

Operations planning procedure

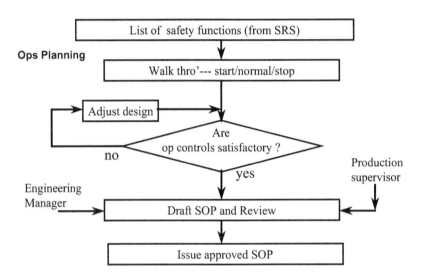

Figure 10.2
Suggested method for developing operating procedures

A suggested general approach to developing operating procedures is outlined in Figure 10.2 and listed below

- Draw up a list of operations required for each safety function.
- Rehearse the operation of each function through the start up, normal running and close down phases of operations.
- Identify any need for operator actions such as reset and start up overrides at each step and ensure the correct facilities have been provided in the design.
- Write up a draft operating procedure and review the details with the persons who are going to be responsible for operation of the equipment or plant.
- Finalise the procedure and issue them in an approved standard operating procedure (SOP) with approvals signed by the engineering manager and the operations manager.

Impact of maintenance procedures on SIS design

We may find that the maintenance work procedure impacts on the design, particularly in the features we build in to assist with maintenance and the control of access to the equipment. For example:

Maintenance procedures define:

- Access control to equipment: e.g. Locks on cabinet doors, flashing indicators
- Facilities for on-line testing such as design and labelling of override switches, power isolation facilities per loop function, layout of terminations, local test facilities for valves and actuators. The design should segregate the input/output terminations and the power isolation to allow one safety loop to be worked on or tested without risk to the others
- Scope of the engineering interface for:
  - Configuration of hardware
  - Access to application software and variables in the data
  - Viewing of resources and diagnostics
  - Copying and backup of application programs
  - Testing of logic solver functions/simulation facilities
- A testing and diagnostic recording facility may be needed to prove that shutdown or isolation valves are functioning

Maintenance planning procedure

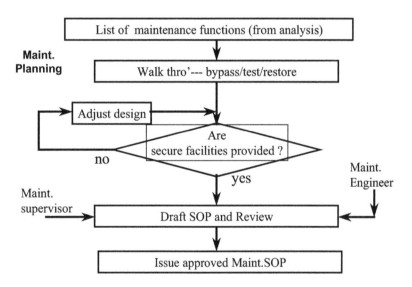

Figure 10.3
Suggested development procedure for maintenance activities

Figure 10.3 shows that the suggested planning procedure is to be repeated for a list of maintenance activities that will lead to an approved maintenance standard operating procedure.

Conclusion

The operations and maintenance planning tasks reveals a lot of things we ought to get right before we go too far. IEC have a good point.

10.3 Validation planning

Validation means providing documented evidence that the installed system meets the original design intent. It is one of the most important ways of ensuring that systematic errors of design have not slipped into the SIS.

Validation is a mandatory requirement for achieving compliance with IEC 51508 as it is an essential element of the safety life cycle. A critical safety instrumented system is likely to be audited with a view to obtaining regulatory or insurance approval to operate.

For all these reasons, the validation stage is a particularly important one.

10.3.1 What should we cover in validation planning?

The validation plan is really a plan to audit the quality assurance trail of the SIS. The main validation task takes place after installation and ideally before critical operation of the SIS. Experience of validation work teaches that the task can be lengthy and difficult to conclude if it has not been done as an ongoing task from the start of design. Catching up on neglected QA trails is called 'retrospective validation' and it is a very inefficient and frustrating thing to have to do!

Validation planning is the task of defining what SIS equipment and what documents are to be verified, what tests are to be done and it includes the task of writing the test procedures.

The way to do the validation plan is to decide the scope of validation work, decide how it can be done in practice and set it down as a list of records, documents and tests that have to be audited by a validation inspector. The plan should also set time milestones for actions so that each audit is done when the information is fresh.

10.3.2 IEC Box 7: overall validation planning

Here is a summary of the scope of requirements based on the details spelt out by IEC 61508 for phase 7 in clause 7.8.2.1

Scope of plan

- Defines when the validation shall be done and by whom
- Specifies the modes of operation of EUC that are applicable to SIS operations. E.g.: Start up, manual mode, auto mode, steady state, shutdown, maintenance, abnormal
- Lists the specifications of the SIS to be validated before commissioning
- Defines the technical strategy to be used (e.g. analytical, statistical)
- Measures, techniques and procedures to confirm the SIS functions are able to perform the required safety functions
- Specific references to each clause of SRS and safety allocations phases of IEC standard (I.e. a per paragraph confirmation)
- Defines environment of validation activities (e.g. test equipment, calibrated tools)
- Pass and fail criteria
- Policies and procedures for evaluating results

Summary

It is clear from the above list that the validation task is a very exacting one. However each subject item in the scope list is not particularly difficult to carry out. If any of the above items presents a problem in a project it may be an indication that the audit trail is not healthy. One might ask, 'Which item would you like to leave out?'

10.4   Installation and commissioning planning

These are practical plans that will link up with the other activities on the plant or equipment being built. Planning will quickly reveal any timing or logistical problems that could compromise the project.

Probably one of the more typical planning problems will be that the state of progress of the rest of the plant will not permit a smooth run for the safety system. For example the installation of instruments on a plant is often delayed by mechanical completion work, hence there is pressure to complete the testing of safety control loops quickly so that the process commissioning can begin.

As we have seen the safety system project requires strong emphasis on QA and validation processes. There is a risk that project time and construction pressures will work against doing the job in a nice methodical and orderly manner. Hence the planning of the installation and commissioning work is a line of defense against these pressures.

10.4.1   What should we cover in installation and commissioning planning?

When we look at the tasks of installation and commissioning we shall see that the planning is quite complicated. The problem will be to synchronise the testing of the SIS with all the construction and commissioning activities of the plant.

The IEC standard does not help very much with the actual details of the installation and commissioning activities, it merely says we must work to the plan! However the ISA standard combined with experience will do the job.

Chapter 11 will map out the installation and testing work. We are going to try starting a plan here and will come back to it with some more details as we go through Chapter 11. Firstly lets check the IEC Phase 8 for advice.

10.5   IEC Phase 8: installation and commissioning planning

Objectives

- To develop a plan for the installation of the SRS in controlled manner
- To develop a plan for the commissioning of the SRS in controlled manner

Installation plan

Needs to include: installation schedule, responsible parties, sequence of integration, criteria for declaring all parts of the PES 'ready for installation' and for declaring installation activities complete. Also requires procedures for resolution of failures.

Commissioning plan

Needs to include; commissioning schedule, responsible parties, procedures, and relationships to installation. Relationships to validation.

Commentary

IEC want us to make sure that we not only know how and when we are going to bring the SIS into service, we need to know who are the responsible parties. We need to define the rules of engagement!

How to develop an installation plan

Probably one of the more typical planning problems will be that the state of progress of the rest of the plant will not permit a smooth run for the safety system. For example the installation of instruments on a plant is often delayed by mechanical completion work, hence there is pressure to complete the testing of safety control loops quickly so that the process commissioning can begin.

Here's a suggested approach to building up a schedule for installation and commissioning of a safety instrumented system.

Activities list

All plans need to start with the activity list followed by the logic or precedence. Each activity can be set down with:

- The details of the procedures to be followed
- The names of those responsible for carrying out the task
- The relevant specification and installation documents
- The criteria for declaring the work completed and inspected

Assuming we have the activities list the precedence logic can be developed as follows:

Step 1: Install sensors

Choose a focal point for the plan…not the start or the end point. The middle is easier, e.g. mechanical completion of the equipment installation or plant. We will have to wait until then before we can safely install the sensors and finish setting up any final elements such as valves.

The figure shown here has the precedence logic resulting from steps 1 to 5.

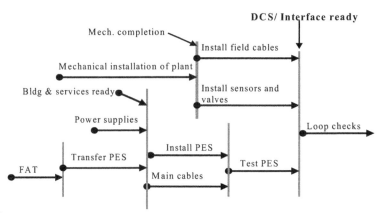

Figure 10.4
Example of part of an installation plan for the SIS

Step 2: Add field cables and do loop functional checks

With sensors installed we will need to do loop checks and calibration tests. Add these activities. Before loop checks we need to run the field cables and terminate securely. Add this cabling work to the plan to start when the PES arrives and finish in time for loop checks.

Step 3: Add the PES

For loop checks we need the PES fully installed and running. Add in the installation of the PES. The PES can function with its own engineering interface, but if the process control system interface is to be used we need a tie in to the installed PLC/Scada or DCS preferably by the time we start loop checks.

Step 4: Add essential services

The PES must be housed in a secure area, normally an equipment room where it can be secure. So we can add in a building and services line to be completed before the PES arrives. It will need stable and proven power supplies so these should be added to be complete soon after the building is ready.

Step 5: Deliver the PES from the vendor or testing site

The PES must be delivered to the site in a fully proven condition. Add to the plan a period for the factory acceptance test and transfer to site when the building is ready.

With the framework in place many other related activities can be added in.

The date lines and the time periods can now be firmed up using the basic planning logic laid down here and so the full plan will develop.

The method is clear; we can come back and do more as the details for tasks emerge in the next chapter.

10.6 Summary

We have seen that there are three distinct and major planning tasks that must be carried out at the point in the safety life cycle where the details of the safety system are well defined. These tasks cover:

- Operations and maintenance procedures
- Overall validation
- Installation and commissioning

Their scopes have been outlined and their impacts on the progress of the safety system have been noted.

Planning can be a tedious and demanding activity but there is wisdom in the IEC requirements that the planning is done in parallel with the detail design.

The designers of the safety life cycle want us to be sure that at all times the design does not lose contact with its objectives, its origins and its final use.

11

Installation and commissioning (IEC phase 12)

11.1   Introduction

This chapter tracks the safety system from its building stage through factory acceptance testing, delivery and installation and into final testing for handover to the operating team. The introduction slide gives an idea of the subjects and the theme...keep track at all times.

11.1.1   Flow chart of activities

The overall flow chart for taking a process plant SIS from design to commissioning is shown in Figure 11.1.

It is normally the responsibility of the project engineering team to take the SIS from the design stage right through to the handover to the operations team or end users. Managing this task requires that continuity be maintained from the design stage.

The best people to test the system and see that it is installed properly are the designers supported by the end user technicians. If this is not practicable the designers should at least specify all the test procedures and installation requirements that are to be observed by the site team.

11.1.2   Procedures

There are a number of well-proven procedures used in these stages that are pretty much the same as we would use for delivering and installing any control system project. We can touch briefly on each stage to get the general idea.

11.1.3   Standards

We will also keep an eye on the mandatory requirements and any additional guidelines described in the standards. As usual the ISA writers are strong on practical experience and have captured their ideas in a simple and concise form in the S84.01 standard.

The requirements in IEC 61508 are simple. Phase 12 calls for the installation work to be done to the plan and for the activities to be documented. There is nothing there to tell us how the job gets done.

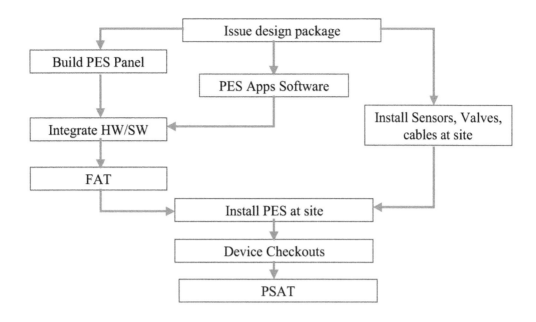

Figure 11.1
Typical installation sequence for the SIS

11.2   Factory acceptance tests

This example of the installation procedure includes factory acceptance testing (usually known as FAT) in the installation phase so that all the testing and proving practices can be seen together.

11.2.1   Scope and benefits of FATs

Before we rush ahead! Let's be clear about the basic steps involved in getting an SIS from the CAD terminal to the final installation.

There are two main types of SIS likely to be seen in the factory test stage:

- Process plant SIS (e.g. chemical plant ESD system)…which has to be integrated with its sensors and final elements at the end user site
- Packaged equipment SIS (e.g. burner management safety system)…which is often fully assembled at the manufacturer's works

The testing procedures for either type are similar but where quantity repeat versions are being produced the potential for a sophisticated testing regime in the factory is greater.

A factory acceptance test allows the PES hardware and the application logic to be tested together at the works where the vendor has assembled the system. In current jargon this would be at the works of the system integrator. If the PES part of the SIS is thoroughly tested with its correct I/O signal channels in service it is reasonable to expect the same results when the instrumentation is linked up to it at the end user's site.

Most companies recognise the benefits of doing the overall system testing at the supplier's works rather than at the site.

The benefits of testing at the works include:

- Resources available
- Avoids stresses found at site
- Provides good training environment
- Brings designer end user and supplier together
- Links testing to a payment milestone

It is important to note the participants for the FAT as shown in Table 11.1
Participants and responsibilities in FAT

Participant	Responsibility
Vendor's engineering representative.	To deliver the finished and qualified PES. Demonstrate in-house system tests.
Design team engineer (Contractor or Client).	Co-ordinate FAT. Accept/fail decisions. Sign off acceptance to deliver.
End user representative. Technician or engineer for maintenance.	Witness, training, application software responsibility. Co-signatory.

Table 11.1
Responsibilities in the factory acceptance test stage

The value of the FAT is that it allows the end user representative to independently test the logic and all other functionality of the PES. If the vendor's engineer has misunderstood anything in the user's specification (i.e. the safety requirements spec) there is a good chance of finding it during the FAT.

11.2.2 Test methods for the FAT

A typical FAT procedure is suggested here:

- The procedures for all tests, physical and functional are defined by the end user or his design contractor in the Factory Acceptance Test Specification (part of IEC phase 8). The results of all tests are recorded in the FAT Report, which will be signed by all participants.
- Physical inspection for correct build standards, cabling practices, terminations and device labeling are done first.
- The serial numbers, model codes and embedded software revision codes are all logged to the test record document. The revision codes of the configuration or application software build are logged.
- Check the relevant testing authority's certificate is applicable to the PES system components with correct version numbers. Review the safety manual and check that all compliance issues listed there have been met.
- Vendor to produce records of the operating systems software build, testing records and configuration parameters. Vendor to demonstrate correct functioning of all diagnostic routines and any automatic protection functions such as 1oo2D switchovers.

- Cross check all hardware and software components against certifying authority certificate.
- Confirm all I/O modules and channel assignments are correctly made and identified in hardware and software.
- Use input test signals and contact closures to identify signals into the PES and check that they are correctly identified in any operator interface. Similarly for output signals.
- Test all signal ranges, under range and over range responses using calibrated instrumentation tools. Test all configured set points for alarms and trip actions.
- Create failure scenarios for CPUs, I/O modules and power supplies.
- Test all non-interactive configured logic. I.e. simple AND/OR logic applications.
- Test the same again in conjunction with bypass or override features.
- Test the specified security functions to prevent unauthorised access to overrides or forcing functions.

11.2.3 Simulation issues

For a fairly simple logic solver unit the above will be sufficient to complete the FAT. However as combinational logic becomes more complicated the potential for errors in the testing methods increases. Often there are more things wrong with the test equipment than there are with the PES logic.

This raises the issue of simulation of the plant signals used to test the SIS functions. There are several methods of creating input/output and application logic testing signals for the SIS application logic.

Testing methods for FAT: 1

Use an internal emulation program within the logic solver to 'play back' the input signals from the state of the output signals. A lockable switchover feature is required to transfer the PES I/O tables over to the emulator section from the physical I/O link.

A separate sub-routine is added to the PES to change related input states when the logic solver changes its output states, thus imitating the regular responses of the plant. The routine responses enable the whole logic configuration to be run in a near normal mode. Test of safety functions can then be easily imposed on the working states of the plant simulation.

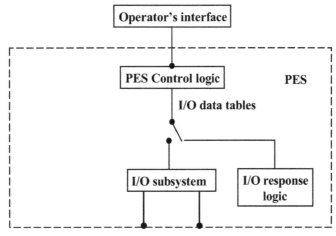

Figure 11.2
Using internal emulation to test SIS logic

One disadvantage of the internal emulator is that it requires the software of the PES to be modified whenever a testing procedure is found to be faulty. Hence there is a risk to the integrity of the overall package.

Testing methods for FAT: 2

Use an external PES with an emulator program to respond to I/O states. This could be the development system (workbench) computer or another PES. If these types are connected via a serial communications link the link must be part of the certified system. This arrangement avoids the need to change anything in the SIS logic solver.

Testing methods for FAT: 3

Use an external PES connected via hardware I/O channels and programmed to respond to output states from the system under test.

Testing methods for FAT: 4

Use a lamp, switch and voltage simulator with manually operated states and responses.

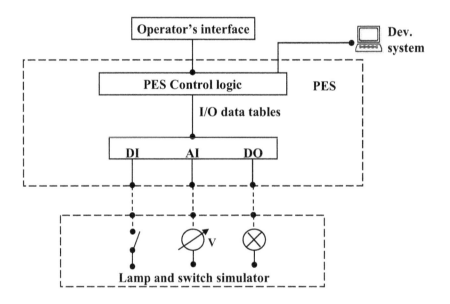

Figure 11.3
Using a lamp and switch simulator to test SIS logic

Figure 11.3 shows the elementary lamp and switch simulator frequently used to test SIS logic. In general sophisticated testing systems are not justified for SIS applications because of the small numbers of system produced and the fact that many applications involve simple logical functions.

This last type is widely used for the final testing in conjunction with the development system serially linked into the SIS logic solver. We should bear in mind that the software development life cycle activities will have utilized emulation of the plant signals during the software integration tests. (Refer to Chapter 10 of the workshop.) These tests will have extensively tested the application logic before it was brought to the FAT. Hence it can be argued that there is no need for complex logical testing at the FAT stage. In such cases the basic lamp and switch simulator will be adequate for the job.

11.2.4 FAT supports functional test specs

Another advantage of factory testing is it provides an opportunity to verify the functional test procedures. These are going to be repeated at the site in the so-called PSAT stage. These same tests form the basis of regular proof testing procedures so the FAT is an ideal opportunity to get these procedures right with the equipment and with the end-user support technicians.

11.2.5 Test facilities in development systems

Safety certified PES products incorporate an approved suite of engineering software tools available in the 'workbench' for managing the design and testing of the PES software.
Features to expect include:

- Revision control of the application:
  This feature compares a new version of the control program in the off-line station with the version installed in the PES. Lists all differences between the two versions and allows verification that the changes have been implemented properly before the new version is downloaded.

- Verification tool
  Uses the above comparison tool to verify that the version of the control program loaded into the PES is exactly the same as the version that has been tested in the off-line station. This supports the IEC 61508 requirement for verification that the correct software has been loaded. Also assists periodic proof testing.

- Live viewing of the logic execution: familiar feature for most controller and PLC development facilities.
- Detailed monitoring of signal behaviour.
- Diagnostic message handling and storage.
- Forcing of input and output signals under password protection with logging of the event.
- Automatic program documentation.

Once the FAT stage has been completed and all documentary records have been tidied up the SIS logic solver is ready for shipment to the site for installation. A relatively simple set of test procedures at site will be used to confirm the unit has been delivered and installed without harm to any of its equipment.

11.3 Installation

With the PES accepted by the end user for shipment it can be taken to the site for installation and linking to the field instrumentation.

11.3.1 Management of the installation phase

Instrument installation contractors are deployed to install all instrumentation in a large plant. The scope covers all field instruments, valves, cabling, power supplies and all control and equipment room hardware.

The installation methods employed for safety instrumented systems are identical to the rest of the instrumentation. There do not appear to be any unusual requirements for site installation beyond the need to ensure the following:

- Segregation from BPCS instruments
- Identification of the safety critical items (e.g. red junction boxes, large red cable markers)
- Provision of on-line proof testing facilities
- Meticulous care to ensure the correct sensors are received, calibrated and installed

Segregation practices typically involve the installation of SIS signal and control loops as completely separate cabling sub systems with their own dedicated cables and junction boxes as shown in Figure 11.4.

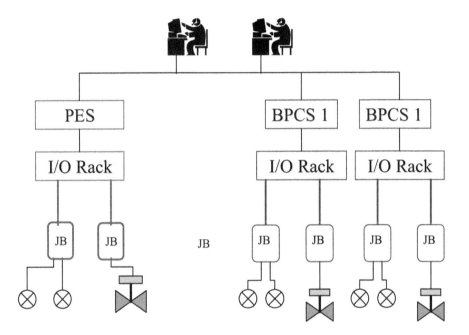

Figure 11.4
Elementary segregation of SIS components

The main purpose of this segregation is to minimize the risk of common cause failures. For example: separate maintenance permits will be used to allow technicians to access the junction boxes for the safety system. This reduces the chances of a routine service job on the BPCS leading to simultaneous interruption of the control signal and safety shutdown signal.

It is a good idea to provide very clear identification of SIS field components by using colour coding (red) and by using clear labels, sometimes with warnings. The fail-safe nature of the SIS loops means that one accidental removal of a fuse or wire link will very likely cause an immediate trip of a large section of plant. Technicians who do this are not popular!

The general installation requirements are all standard practice for a good quality installation contract. Most large operating companies have established codes of practice for instrumentation installations that seek to:

- Standardize designs and define the exact components to be used

- Prevent the use of poor quality fittings
- Provide a consistent standard for training of technicians

11.3.2   Installation checks

The checkout of completed installations is normally carried out in two stages exactly as performed for a normal control system. The stages are:

- Device physical checkout
- Device functional checkout

Although this is standard installation practice for instrumentation the installation stage always presents opportunities for the various parties to get it wrong. One reason for this is the interface between instrumentation, mechanical and electrical teams.

Check team

The checkout should be performed by a team comprising:

- The contractor's technician
- The design engineering representative
- The maintenance technician

All participants are expected to sign off each checkout form at the end of testing and all comments; especially 'reservations' will be documented. A dispute resolution procedure is essential for this stage as was noted in Chapter 10 when we covered the planning phase.

Physical checkout items should show details for the mechanical installation such as the simple matter of identifying that the right valve has been installed in the line. Since so many valves look alike it's easy to get a near miss! Similarly the details of impulse tubing can be critical to the sensor operation whilst the contractor sometimes finds scope for some innovation in the way the lines are arranged!

Physical checkout package

An effective way to ensure an accurate record of the installation checkout is to make up a composite package of data sheets and records for each field device. For example a copy of the sensor instrument specification can be clipped to the device checkout form for each safety function as shown in Figure 11.5.

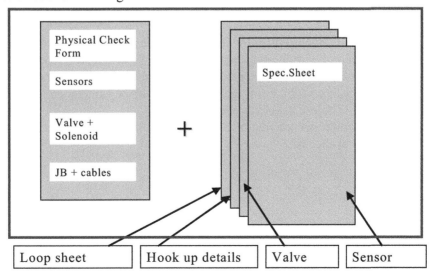

Figure 11.5
Document pack for physical checkout

We can add the process hook up diagram (this details the tubing and valve connections from process to transmitter) and the loop diagram (this details the circuit connections resulting from the correct termination of all cables). Now the checkout form will have all the valid reference data attached to it for a permanent record folder.

The value of this is:

- The checkout team has convenient access to the relevant data sheets at the time of inspection
- Copies of the data sheets valid at the time of the checkout are kept as a permanent history record of the loop along with the signed off check sheets

Functional checkout package

The list here indicates the scope of the functional checkout.

For each sensor: witness and sign off:

- Process simulation signal injected via testing port or tested by physical condition
- Demonstrate calibrated range readout at PES interface
- Demonstrate alarm and trip set point crossing actions
- Demonstrate all bypass functions and associated warning idicators
- Demonstrate security functions for input test overrides

For each final element: witness and sign off:

- Test valve trip actions, directions and speeds
- Record response times and verify against specifications
- Test position feedback functions
- Test drive trip/interlock functions and feedbacks
- Test failure mode actions on loss of power and/or air
- Demonstrate all latch/reset functions
- Demonstrate security functions for output forcing

The device functional checkout should be used to test all individual functions of the sensors, final elements, the operator interface and the maintenance interface. This is the best opportunity to rectify any defects that may have crept into the detail design or the installation. At this stage it troubles nobody else if a malfunction is discovered, the contractor is on hand to rectify problems and it may be some time before the final pre-start up acceptance test is to be conducted. Hence get as much right as possible before the next stage.

A similar documentation arrangement can be used for the functional checkout as for the physical checks.

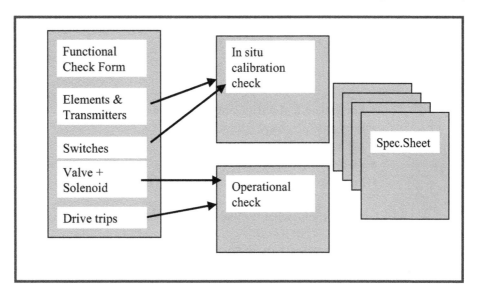

Figure 11.6
Document pack for functional checkout

It is not unusual at this point to uncover something quite silly. For example a control valve may have its springs reversed such that it trips in exactly the wrong direction. Or there may be a combination of tubing and springs that appears to work correctly until the air supply failure test is applied.

For large tripping devices such as large ESD valves on pipeline duties it is good practice to record the basic position versus time response of the actuated valve. Sometimes constricted venting of the actuator causes excessively slow trip response and an improved venting arrangement must be fitted.

Firstly the response must be checked to see if it is within the time specified by the SRS. Secondly the record of this response under no load conditions can be placed into the maintenance record to assist with performance trend monitoring. If you are lucky enough to have one of the new smart actuators this where you can set the response standard for the self-test functions.

11.3.3   Installation complete

With all device checkouts done the installation phase will be considered complete apart from any agreed exceptions or 'reservations'… an unpopular concept!

At this point the installation contractor normally gets paid for a major part of the work, hence the pressure to complete the checkouts. The user representative has maximum leverage up to this point!

From here on it has to be assumed that all devices in the SIS including the logic solver and operator interface are fully functional and correctly installed. This provides a firm basis for the PSAT to go ahead.

ISA requirements for installation and commissioning

ISA has this to say about basic installation work:
'8.2 Installation
8.2.1 All equipment shall be installed per the design'

And then goes on to provide clear requirements for testing once the basic installation work has been done.

.............confirmation that the following are installed per the detailed design documents and are performing as specified in the safety requirement specifications:

- Equipment and wiring are properly installed
- Energy sources are operational
- All instruments have been properly calibrated
- Field devices are operational
- Logic solver and input/output are operational

Which is effectively what we have done in the checkout stages.

11.3.4 Pre-start-up acceptance tests (PSAT)

Clause 8.4 of ISA S84 then goes on to describe the PSAT, which is a full functional test of the complete safety system.

PSAT provides a full functional test before introduction of hazardous materials. Includes:

- SIS communications to external networks or BPCS
- SIS components perform to SRS
- Trip set points
- Proper shutdown sequence
- Annunciation and displayAccuracy of any computations
- Reset functions
- Bypass and bypass reset functions are correct
- Manual shutdown systems operate correctly
- Proof test intervals documented in maintenance procedures
- Proof test intervals consistent with SIL calculations
- SIS documentation consistent with install/operate procedures
- Accuracy of calibration equipment consistent with resolution/accuracy of application
- PSAT report to be completed before PSSR
- PSSR must be done before hazards introduced. Authorized signatures to be completed.

As noted in the planning stages all details of the specific PSAT should have been set down during the design stage by the design team. The design intent will then have been captured in the safest way.

Both ISA and IEC make it clear that all PSATs must be carried out, completed and validated without any process fluids or materials in the plant since this is the final stage before the process commissioning takes place.

11.3.5 Documentation for the PSAT

In addition to the PSAT test procedures we have mentioned the complete set of documents used for the SIS detail design should now be available to support the PSAT documentary records. The list of documents is described in Appendix A.

11.3.6　Validation

The task of validation described in the IEC safety life cycle should now be carried out. If the documents listed in Appendix A are available and all the revisions are clearly up to date and recorded in a register, the validation process will be relatively easy. A quick glance at the IEC model should tell us where we are.

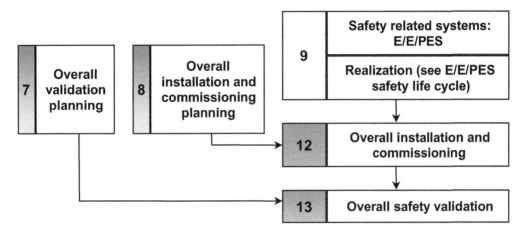

Figure 11.7
IEC installation and validation phases

An examination of the contents of the IEC validation clause reveals that it requires the documents described above and the PSAT also as described above.

So the validation work is really a carefully documented version of the PSAT. It is work that should be done by an independent auditor who can study the records impartially and be satisfied that no items remain unclear or untested.

With this stage completed we are nearly ready for the operation of the SIS to commence. The next stage is likely to be start up of the process or machinery. This is likely to be the stage where the design of the SIS is really challenged since the real physical conditions will now be imposed on the sensors and final elements. So we cannot assume that the job is finished. There will inevitably be adjustments to the SIS and possibly some logic changes to be made when the real plant starts to run.

11.3.7　Training of technicians and operators

Before we move into start up and operations the issue of training must be considered.

Commitment to training on the safety systems should be made from the outset of the plant project. It is therefore good policy to involve the end user technicians in training from the moment the SRS becomes established in the design phase. If the key individuals can be involved in all subsequent stages of the life cycle it will be of great benefit to the success of the installation and its continuing upkeep.

The key points to note here are:

- Involve plant technicians as soon as the SRS is issued
- Identify key support individuals: plant and vendor
- Key individuals to participate in:
    - Programming (configuration) of the PES
    - Preparation of FAT/PSAT procedures

- – Field and Panel location choices, accessibility issues
- – Preparation of maintenance procedures
- – Training manuals
- – FAT/PSAT/Start up

The American OHSA regulations define requirements for training with respect to safety systems. IEC requirements for training are based on IEC clause 6.2.1 (h) supplemented by Annex B concerning the Competence of Persons. These requirements specifically mention the need for the training of staff in diagnosing and repairing faults and in system testing.

11.3.8   Handover to operations

The key points at the handover stage are:

- The operations team are responsible for SIS from start up
- Validation responsibilities to be agreed
- Handover accepted at the end of PSAT, provided:
    - – There are minor reservations only
    - – The documentation package is complete
    - – Training competence has been established

The normal practice in the process industries is for the operations staff to take over ownership responsibilities for all instrumentation and control systems at the start of process commissioning. i.e. before process materials are introduced to the plant. This allows all plant safety procedures to be enforced from the moment there is any potential for hazards due to the process materials.

It is clear that the operations team must be satisfied that the SIS is operational and does not require any significant further work to be done on it. The completion of the PSAT therefore marks the point where the handover of the SIS from the project team to the operations team should take place. Operations will not accept responsibility for the system if there are any reservations that would prejudice the safe operation and maintenance of the SIS.

11.3.9   Start up

Start up is potentially the most hazardous phase in the life of the plant because:
- The equipment is new and untried
- Experience is limited
- Abnormal operating conditions (e.g. stage bypasses) will be used until all units have been properly commissioned
- Stress levels are high

On the other hand there is a greater than normal concentration of engineers, technicians and managers all watching out for problems! The real hazards may occur later if familiarity allows good procedures and practices to fall away.

There is clear need for the SIS maintenance and operating procedures to be carefully observed during this phase. The potential for quick fixes is high. This is where the benefits of responsible ownership of the SIS by the plant personnel will be realized.

11.4 Summary

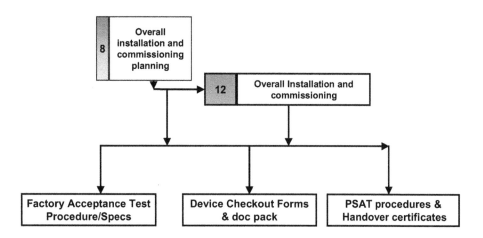

Figure 11.8
IEC installation phases

Figure 11.8 shows the key features of the installation phase. We have taken a walkthrough of the stages right up to the start up of the plant. The picture is one of methodical checkouts and documentation records at each step. The activities are supported each stage by the end user technician where possible or available.

This idealized picture is by no means the way it actually happens and we must be aware that circumstances often conspire to destroy this neatly ordered chain of events!

The relevance of the IEC planning phase in preparing carefully for these stages can now be seen clearly. If we can at least lay down all the tests we require and the procedures to be followed these will serve us well when things threaten to go off course.

11.5 Documentation required for the pre-start up acceptance test

Documentation for the safety life cycle will hopefully be cataloged throughout the life cycle of the project. From the listing, the following documents should be made available at the time of the PSAT:

1. PSAT procedures as developed during phase 7 of the SLC
2. Copy of the SIS safety requirements specification
3. Block diagram of the overall SIS sensors, logic solver actuators, interfaces, displays/alarms, BPCS interfaces etc
4. References to P&IDs (separate copies not recommended)
5. I/O list with links to the instrument index
6. Copy of instrument index
7. Specification sheets for all main components including sensors and actuators

8. Loop diagrams
9. Electrical interface schematics
10. DCS configuration or instrument philosophy diagrams for any loops involving DCS inputs or outputs
11. Instrument and panel location drawings
12. References to cabling schematics and junction box listings
13. Pneumatic hook up diagrams or cross references to pneumatic hook up standards
14. Register of vendor's product documentation and manuals with access to all items
15. Trip logic or cause/effect matrix diagrams for the logic functions
16. PES application software sheets or functional logic diagrams PES programming documentation hard copies
17. Copies of safety certification for SIS logic solver components and any available for sensors
18. Qualification files for any uncertified sensors and actuators.

12

Validation, operations and management of change (IEC phases 13, 14 and 15)

12.1   Introduction

Before we move on to maintenance and functional safety testing we need to deal with the subject of 'functional safety assessment'. As we know IEC 61508 is very keen on maintaining a continuing process of crosschecking to see that the objectives of any activity in the SLC are properly met.

This chapter starts by sorting out the difference between verification and validation. It then examines the meaning of functional safety assessment. Then we take a look at the operating viewpoints on the safety system, its maintenance regime and finally consider the subject of change management.

12.2   Verification, validation and functional safety assessment

We have lumped the three terms 'verification, validation and functional safety assessment' into a package so that we can see how they compare and can clarify the differences between the terms.

Let's first examine what these three terms mean.

12.2.1   Verification

Dictionary:

Verify = 'establish the truth or correctness of by examination or demonstration.'

Verification as per clause 7.18 of IEC 61508-1:

Objective: '... to demonstrate for each phase... (by review, analysis or tests) that the outputs meet in all respects the objectives and requirements specified for each phase.'

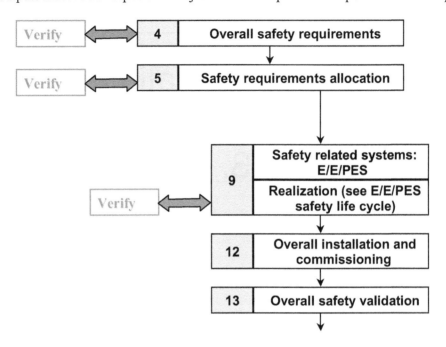

Figure 12.1
Verification takes place at each phase of the SLC

Basically verification is a QA check on each phase of the SLC to see that it has achieved its objectives and met the requirements. Clause 7.15 of IEC requires that a plan for verification be established concurrently for each phase. For example:

Verification: example plan and review

In the allocation phase of the high-level protection systems example there are 2 objectives to the phase:

- Allocate safety functions to a designated E/E/PES safety related system
- Allocate an SIL to each safety function

There are also 13 requirements sub-clauses: 7.6.2.1 to 13.

A plan for verification might look like this:

At the end of the work in this phase complete the following checklist:

Item	Question	Y	No	Comment
1	Have all safety functions been allocated	√		Only one
2	Has SIL been allocated to each function	√		
3	Have designated systems been specified	√		PES No 1
4	Have skills and resources been considered	√		Std tech skills
5	All safety functions allocated and met	√		Flameproof motor + fence
6	Is target parameter PFDavg or λd	√	√	
7	Has the allocation been based on a probability basis?	√		
8	Has common cause failure possibility been considered? State the independence factors	√		Level detector system and final element all separate from BPCS
9	Confirm the SILs for each function	√		SIL = 1 for level No other functions
10	Does PES serve multiple safety functions If yes what is the highest SIL Does PES meet the highest SIL	√	√	
11	Confirm the SIL is less than 4	√		
12	Has all information and results for this phase been documented into the files for this process What is the document ref no and revision date Has it been signed off by the authors?	√		Not yet

Table 12.1
Checklist designed for verification of phase 5 of the SLC

Note that by making the above checklist as part of the verification plan we have arrived at a checklist that can be related back to the conceptual design phase.

So verification is a matter of good QA practice applied at each phase of the SLC. This ensures consistent carry through of the intentions of the safety requirements spec.

12.2.2   Validation

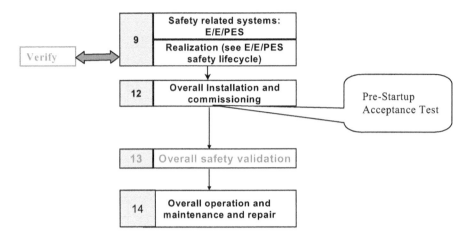

Figure 12.2
Validation is completed after the commissioning phase

The role of validation is to prove that the end result matches the original intent
Dictionary:
Valid = sound, defensible, well grounded, executed with proper formalities.

The previous chapter has shown that validation is carried out at the end of the installation phase to provide documented evidence that the system meets its original requirements, (i.e. defensible, executed with proper formalities). Validation comprises a review of the documentation and testing records for evidence that the safety requirements have been met.

A simple comparison with verification may be useful here:

- Verification shows that the specification and design and build stages have progressed in a correct manner
- Validation shows that the end product meets the original requirements

12.2.3   Functional safety assessment

There seems to be a lot of duplication in the validation type of activities but in practice each serves a different purpose. Functional Safety Assessment (FSA) is defined in a separate clause in IEC, clause 8.

Objective: '… to investigate and arrive at a judgement on the functional safety achieved by the E/E/PES safety related system.'

Here is a very brief summary of the requirements for FSA based on the details set out in clause 8 of IEC 61508-1.

Requirements: (Clause 8)

- FSA to be planned and agreed with all parties
- One or more persons appointed
- Access to all persons and information
- FSA applied to all phases of the SLC
- To be done before hazards present
- Tools used are part of the FSA
- Recommendations to be produced

It should be apparent from the above description that the FSA is a critical appraisal of the functional safety of a given process plant or boiler station for example. The appraisal is achieved by a continuing process of reviewing the activities carried out during each phase of the safety life cycle and by examining the outputs of each stage to see that they meet the requirements of the standard.

The method is similar to an ISO 9001 audit and is likely to be used in the future for linking to insurance company assessments and operating permits for plants having the potential for serious accidents.

Graded independence of assessors

In line with all the other measures mandated by IEC 61508 the scope of FSA work is graded to match the level of safety integrity required of each particular application. The levels of competence and independence for assessors are required to be in proportion to the severity of consequences of the possible accident.

This issue raises the question of where do we find such assessors? They have to be competent and they have to have some level of independence. Notice that for lower SIL levels the persons can be drawn from within an end user organization. Hence the company could train its own people for the job but for a high SIL application a third party is needed.

CASS developments

Anticipating the requirement for FSA services the established certifying organizations already offer FSA services. In the UK a new company has been formed to specialize in providing training for conformity assessment and has taken the name 'The CASS Scheme Ltd'. The implications of this scheme are that competency guidelines and training schemes are in place for persons to be trained as Functional Safety Assessors.

In summary the result of the step-by-step verification of life cycle phases leading to final validation is combined with continuing assessment of functional safety practices to deliver the complete QA package for the safety system. Implementation of the overall scheme will require considerable diligence and discipline.

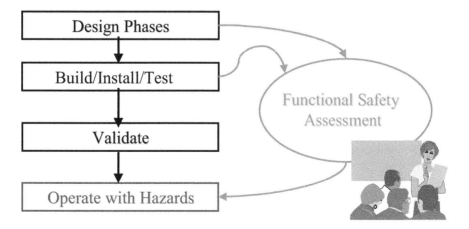

Figure 12.3
The role of FSA is to evaluate the effectiveness of all stages of the SLC

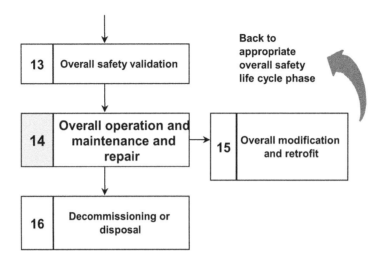

Figure 12.4
Phase 14 operations, maintenance and repair

12.3   Operations, maintenance and repair

So we finally get to use our SIS in beneficial service!

This stage is IEC phase 14 as shown in Figure 12.4. The standards have very little to say about the routine operation of an SIS in terms of operating procedure. This is perhaps because it is normally a passive system that remains unobtrusive and does very little until called upon to act swiftly in a hazard demand situation. It may however be useful at this stage to review the typical features of an SIS as seen from an operator's viewpoint.

12.3.1   Operator's viewpoint

The characteristics of a typical SIS can be listed from a process operator's viewpoint:

SIS status indicators

Indications on the BPCS workstation or the control panel that all SIS system functions, self-tests, power supplies etc are healthy. Prominent alarms will display if an SIS system fault occurs, messages will indicate follow up actions urgently required.

Process indications

Indications on the panel or workstations of the process values being transmitted by the SIS sensors will normally be available.

Permissives

During start up of a process unit operation there may be indicators to show that the SIS cannot permit start up until certain conditions have been met or reset. As these clear down the next step in the start up can be followed. These indicators are sometimes known as 'permissives' and are used as milestones in a start up sequence.

However the SIS logic should not be used for direct control of start up progress, rather the main BPCS should be programmed to manage all start sequences such that blocking by the SIS should not have to occur. Remember that it is a basic principle of SIS design that it should not have to play any role in routine control operations. For this reason an SIS trip should not be used as a routine means of shutting down a plant unit.

Status of Drives

It is common practice for the trip and interlock status of individual drives to be shown on a BPCS display. Hence the reason why a drive may not be available to start is shown and this assists the operator with diagnosis of an operational problem.

Process overrides

Some start up conditions may have to have a 'process start-up override' condition set by the operator. These conditions will be indicated by a flashing indicator to remind users not to leave things in this condition. The safest practice is provide a timeout function in the SIS logic so that the override removes itself after a practical interval.

SIS alarms

The SIS will normally have warning alarm functions provided, sometimes called pre-trip alarms, warning of an approach to a limiting trip condition. The operator will be alerted to this as a high priority alarm demanding urgent remedial action.

Trip alarms

When a trip condition occurs the SIS action must always be programmed to create a trip alarm indication to the operator. A golden rule in any design, however small is 'NO BLIND TRIPS'.

Alarm and event logs

The SIS will preferably incorporate a sequence of events log file, which can be inspected after the event to diagnoses the cause of a trip. These SOE functions require very accurate time stamping of the events to allow proper analysis of 'what happened first'.

Trip resets

Many trips have the characteristic of being liable to return on their own to the running condition when the problem goes away. In many but not all circumstances this should not be allowed to happen, hence the use of a latching trip logic stage in the SIS or a mechanical latch at the solenoid valve. From an operator's viewpoint the latches must be reset before the plant can be restarted. If the latches are programmed into the logic solver then a reset control button will be needed for the operator to use. Sometimes these will need to be key operated to reduce the abuse of this facility.

Permits for maintenance work

Whenever any maintenance is in progress on an SIS the work will be first authorized by a permit to work or 'clearance'. All operators are routinely made aware of such a permit in operation through a prominent display board or book in their control room. The shift supervisor will normally be one of the signatories to the permit.

Maintenance bypasses

When any maintenance bypasses or overrides have been applied the SIS circuitry must be arranged to provide a flashing warning indicator in the control room. Operators must always be aware of the particular override in force. Therefore, a specific alarm indicator supports the flashing indicator or screen message telling the operator which device is on override and which safety function(s) is in a maintenance mode.

Variable set points

In some types of processes and particularly in batch processes the situation may arise where the safety limits of a process will change according to the state of the process. An explosive condition in one stage may be safe and allowable at another time. SIS designs may, in these cases, allow the operator or a program in the BPCS to change the set point of a trip system within certain permitted boundaries. Clear operating instructions and good training manuals are obviously essential in such cases. The preference would appear to be for the SIS to derive the trip points dynamically according to a set of variables it is able to measure. In this case the protection task will remain invisible to the operator except for an indicator tracking the derived set point.

12.3.2   ISA requirements for operating procedures

There are some mandatory requirements given in ISA S84.01 in:
Clause 9.4 'SIS operating procedures'
Summary of ISA SIS Operating Procedures

- Procedures to be written to explain safe and correct methods
- Typically part of unit operating procedures
- Explain the limits of safe operation (I.e. trip points), consequences
- How the safe takes the unit to a safe state
- Correct use of bypasses, permissives, resets, etc
- The correct operating responses to SIS alarms and trips.

12.3.3   Maintenance program

ISA requirements

ISA in clause 9.5 has some basic requirements for the maintenance program, i.e.
- Regularly scheduled functional testing
- Regularly scheduled preventative maintenance, filters, batteries, calibration, lubrication etc
- Repair of detected faults with appropriate testing after repair

IEC requirements

IEC has some interesting activity models in Phase 14, which are useful to examine. Figures 12.7 and 12.8 are based on diagrams in Figure 7 of IEC 61508-1.

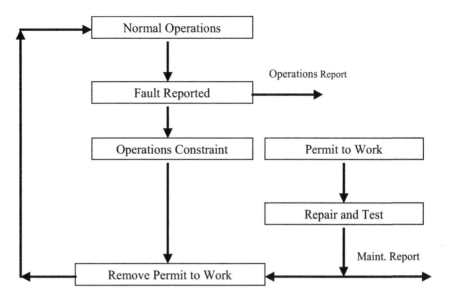

Fig 12.5
Revealed faults procedure

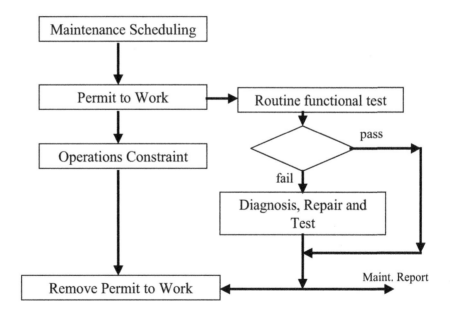

Figure 12.6
Unrevealed faults procedure

The schemes for revising the design in response to analysis of faults or deviations in demand rate is given in Figure 8 of IEC 61508-1 and an outline version is shown in Figure 12.9.

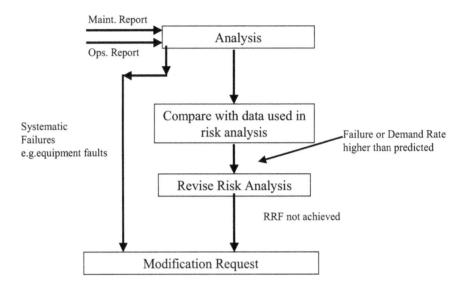

Figure 12.7
Operations and maintenance management model

Given the above models the framework for a management system can be seen:

- Record all events and failures
- Analyze failures and demand rates
- Compare with design assumptions
- Decide on modification request

The decision to request a modification then launches the activities back into the safety life cycle and follows phase 15 of the IEC model. We will visit that briefly in a while, but first we look at functional testing.

12.4   Functional testing

12.4.1   Why test?

- The only way to find unrevealed faults
- PFDavg is reduced and kept within SIL targets by functional testing
- OHSA requirements
- ISA S84 mandatory Section 9.7
- IEC 61508-1 clause 6.2.1...k and clause 7.15.2.2

Paul Gruhn and Harry Cheddie comment in their book that 'it's not unusual to encounter safety systems in plants that have never been tested after installation'. This gives an indication of the present gap between the theory of safety management and real practices.

Put simply: PFDavg increases in proportion to the test interval, SIL falls to level 0.

12.4.2 Testing guidelines

Clause 9 of ISA S 84 provides a good list of requirements and ground rules for functional testing and some of the key points are summarized here. For fuller details it is a good idea to consult the ISA standard.

Test interval

In ISA S84, annex B 15 spells out the ground rules for determining the test intervals:

- Manufacturer's recommended test interval
- Good practices
- As needed to meet the SIL

It is not usually difficult to arrive at the figure for the test interval if the reliability analysis has been done quantitatively at the conceptual design stage. Most operating companies will have default intervals of 3 or 6 months, which will serve as a trial value for a PFDavg calculation.

e.g. for an instrument with an MTBF of 10 years.

$$\text{PFD avg} = (0.1\,\text{failures/yr} \times \text{TI/2})$$

if TI is 6 months,

$$\text{PFD avg} = 0.1 \times 0.5/2 = 0.025 = 2.5 \times 10^{-2}$$

which fits for a SIL 1 mid-range PFD avg.

It's easy to turn the calculation round and extract the TI value for a target PFDavg. The problem is that the answer needs to be kept inside a practicable range. However the test intervals must calculated on the basis of the complete safety function and must be checked against the original assumptions recorded in the conceptual design phase.

Practical situation for testing at a typical plant

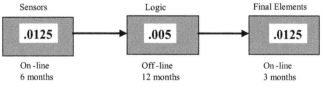

$$
\begin{aligned}
\mathbf{PFD_{avg}} \quad &= \ 0.5\,(0.05 \times 0.5) + 0.5\ (0.01 \times 1) + 0.5\,(0.1 \times 0.25) \\
&= \ .0125 + .005 + .0125 \\
&= \ .03 \ \text{or } 3 \times 10E -2 \quad \text{in SIL 1 range}
\end{aligned}
$$

Figure 12.8
Example, practical scheme for SIS testing intervals

A typical scenario for a complete SIS will be to carry out on-line testing of the sensors perhaps every 6 months because the MTBF figures are fairly low and the PFDavg must be kept to a low figure through the use of a short testing interval. The final elements if they are valves may be tested by partial closure every 3 months due to their low PFDavg values.

The logic solver however may have a much higher MTBF and hence its PFDavg will be perhaps in the $10E^{-3}$ region. The plant has an annual 1 week shutdown for major maintenance so the logic solver can be given a TI test interval of 1 year and the testing can be done in off-line mode when a full functional test of the logic responses can be done whilst the plant is down.

12.4.3   Practical functional testing

Sensor test overrides for sensor testing on-line

To test any sensor on-line we need to block the tripping action in the logic solver whilst exercising and proving all other functionality. The sensor test will need to include the following tasks:

- Block (or override) the effects of the sensor trip
- Alert the operator of an override in place
- Operate the sensor over its range
- Confirm the input tripping alarm and any pre-trip alarms
- Record the values at the trip point
- Adjust set-points if necessary
- Check response rate where critical

Figure 12.9 shows a generic input arrangement for to test overrides to meet the above requirements.

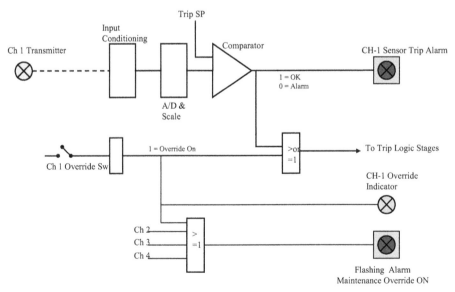

Figure 12.9
Generic input arrangement for test overrides...analog

The arrangement tests the complete input and comparison stages of the SIS. Some PES based systems may prefer to use forcing of inputs to test the comparison function and provide a separate means of reading the sensor range. In all cases the requirements to keep the control room or panel alerted to the presence of an override is of prime importance.

The arrangement for digital contact inputs is identical except that the signal scaling and comparison stages are omitted.

Output overrides

Output overrides are commonly used on relay based safety systems to provide a means of on-line testing the relay logic functions in the absence of any diagnostics. Tripping functions can be demonstrated right up to the final tripping relay where the output contacts can be bridged across by a test override. This practice is risky and has fallen away as PES and solid-state systems brought in diagnostics for short pulse testing of output stages.

Practicalities of on-line testing

Most pressure sensors can be tested on-line if they are fitted with a means of isolation from the process and branch for injecting test pressures for test calibrator.

For testing of other sensors the techniques are not always so simple or effective. Here are a few suggestions. The workshop attendees may have their own ideas to add here. With the aid of the flip chart we will sketch out and discuss methods for typical sensors. These will include:

- Temperature testing as marked ... simulation at the transmitter
- Temperature testing as marked ... hot well calibrator
- Flow meter testing as marked ... simulation at the transmitter
- A less than adequate proof test
- Level testing as marked ... DP type: simulation at the transmitter
- Level testing as marked ... nucleonic type: simulation by shuttering
- Level testing as marked ... float chambers

Every application has to be considered on merit and the method must be documented and agreed for each SIS. Two basic rules:
- The method must not take a long time since override time adds to PFDavg
- The method must not carry a risk of leaving the instrument not working

Since many instruments now include integral manual test facilities the selection of a suitable type may be influenced by the validity of the available testing method.

Testing of final elements

Most final element testing, particularly of electrical drives is best carried out when a scheduled stoppage of the plant is expected. Simulated testing is unlikely to create a realistic test other than proving the ability to de-energize the circuits up to the final tripping relay, which will be bypassed.

Trip testing of final elements up to and including the solenoid valve is useful since the reliabilities of solenoids are variable according to quality and environmental conditions. Also the possibilities of wrong connections of cables and pneumatic tubing are quite high due sometimes to the multiple numbers of batch and control valves present around a reactor for example.

Partial tripping of shutdown valves can be considered where the travel time is reasonably long but of course latching solenoids cannot be used in such applications.

Testing of ESD pipeline valves

Pipeline EDS valves are installed specifically to ensure isolation of an offshore installation from the pipeline risers carrying flammable oils and gases. A high level of assurance is needed that the valves will close fully when they are tripped. At the same time oil and gas platforms are desired to be kept running around the clock for long periods.

The use of a bypass is not permitted since this adds to the risks of failure of joints and valves. In any case the use of a bypass to assist trip testing creates a potentially false result since the supply line pressure is not imposed on the closing valve. It is ability of the ESDV to close against full flow of the pipeline under pressure that is one of its essential features.

The typical practice for inspection and testing of these valves is:

- Physical inspection for leaks, mechanical damage and corrosion every three months
- Partial closure tests every six months (valves have a slow rate of travel)
- Full on-line closure tests every 12 months

It also good practice for these valves to be fitted with position transmitters so that a chart record can be made of the displacement versus time responses of the valve stem. This characteristic time/position plot assists in predictive maintenance and evaluation of the valve's trend in performance. This technique has potential use in many other large valve applications.

Recording the functional tests

The trip testing work will be of limited value unless it is done to a consistent method and is consistently reported. When this is done the records will support the good safety management practices of the plant. The data available from the test records will assist with the evaluation of performance that we have seen must be done periodically as required by the standards.

The best way to ensure the recording of the testing is done consistently is to set up a 'trip test procedure' sheet for each safety function and to include spaces for the regular data such as test values and times to be filled in. Filing of these reports in a secure manner is essential and if a PC database system is used for the purpose it will further enhance the quality of the record keeping. Indications are that a maintenance software package is the ideal tool for this purpose.

12.5   Management of change

Figure 12.10
Why we need to manage all changes

The need to manage change is well understood in respect of safety systems as can be seen by the allocation of phase 15 of the IEC safety life cycle to the subject.

The notorious Flixborough disaster in UK came about as result of a change in plant configuration that went horribly wrong. In that case a series of 6 large reactors, each holding 20 tonnes of cyclohexane derivatives were normally linked in a cascade arrangement via a flexible joint. When one of the units developed a leak the production team decided to organize a temporary bypass of the unit with a pipe made up at plant apparently on the basis of a chalk drawing marked on the floor of the workshop. The pipeline was only tested to a pressure of 9 bar instead of the unit's relief valve pressure of 11 bar.

The temporary pipeline operated for two months until a slight pressure rise occurred and caused the temporary pipe to twist and rupture the flexible connection leaving two 700mm pipe openings to atmosphere from the working reactors. 28 people died in the resulting explosion and the plant was destroyed.

In the Flixborough case the change that was not managed was an equipment design modification. The procedures for the change were not followed and the design change was not carried out by qualified personnel. It was not a direct modification to any safety instrumented system. However as we have seen at the start of this workshop the process and equipment design and its functional safety are all part of an overall quality assurance philosophy for safety. Any change in one discipline must be tested against the knowledge of the others.

For functional safety systems the risk is that instrumented systems can be affected by changes in mechanical, chemical or electrical equipment that they are serving. It is particularly important therefore that strict change procedures be put in place for the SIS.

12.5.1   When is MOC required?

ISA S84.01 has mandatory requirements for management of change. These are set out in clause 10 the standard. It calls for a written procedure to be in place to: initiate, document, review the change and approve changes to the SIS other than 'replacement in kind'. Here is a summary of some of the possible reasons for change that could give rise to a MOC procedure.

Reasons for invoking an MOC procedure

Modifications proposed for:
- Operating procedures
- Safety legislation
- Process design
- Safety requirements specification
- Corrections to software or firmware due to errors
- To correct systematic failures
- Due to a higher failure rate than desired
- Due to increased demand rate on the SIS
- Software revisions: embedded, utility or application

12.5.2   When is MOC not required?

Obviously for routine replacements and repairs an MOC procedure is not required but beware that a replacement must be exact 'like for like' and not a 'generic' equivalent. For a 'generic' replacement, the potential for a changed failure mode or revised reliability figure would have to be evaluated through a change procedure.

Changes falling within the range of adjustments permitted by the SRS would not require a change procedure. Another good reason to do a good job of the SRS!

12.5.3   IEC modifications' procedure model

IEC Phase 15 has a particularly useful model for the activities of MOC and this can be found in section 7.16 of IEC 61508 part 1. Details of the change procedures required for conformity to the standard are spelt out in this section. The key points are:
- Modification requests may arise for numerous performance reasons including process changes and operating experience.
- Impact analysis is applied to each request with a close check on the data held in records of the original hazard and risk analysis. In some cases a revised hazop study may be needed.
- The impact analysis is reported to a modifications log file and referred to the responsible person for authorization. The authorization, if granted is recorded in the log file.
- The appropriate phases of the safety life cycle are then activated and updated to implement the change in the same way as new SIS design.

12.5.4   Impact analysis

The impact of a change to the SIS must be considered and recorded before proceeding to implement a change. A list of impact possibilities can be used for this such as those given by ISA S84.01 and listed here:

- Technical basis for proposed change
- Impact on safety and health
- Mods to operating procedures
- Time needed for the change
- Authorization requirements
- Availability of memory in the PES
- Effect on response time
- Change method: on-line or off-line and the risks involved.

ISA goes on the require that;
- The required safety integrity has been maintained
- Personnel from appropriate disciplines have been included in the review process.

This is effectively the 'impact analysis' called for in the IEC model.

12.5.5 Software changes

Changes to software are a popular activity. The problem is that they seem small and are easy to make but can have a far-reaching effect. The golden rule for the SIS logic solver is that 'Any changes to software should require re-testing of the logic'. This applies particularly to the revisions to the system software made by the vendors. Does migration to new revision of the system require the application logic to be re-tested? I would say yes.

Concessions for logic changes

At the application level there are more software tools available now that allow code comparisons to be made between a revised copy of the program and the earlier version so that very small changes can be validated without major re-testing. This should still not be done to a validated system without going through the MOC procedure.

12.5.6 MOC Summary

Managing changes is essential. The potential for shortcuts is high and the temptation to take them is strong. An agreed and enforced MOC procedure is needed to make sure all disciplines work to the same rules.

12.6 Summary

In this chapter we have seen a wide range of activities associated with the operation and continuing use of the SIS.

The basis of the working SIS is the verification of the design and the validation that it meets design intent, supported by an independent FSA. Operations follow the procedures and testing rules laid down in the design and maintain strict control over changes.

13

Justification for a safety instrumented system

13.1   Introduction

The safety life cycle model addresses the overall and detailed functional safety of a process or equipment system. It assumes that an appropriate decision will be made on the level of risk reduction to be provided by both SIS and non SIS solutions, usually working in some combination to achieve the target risk reduction.

Once the need for risk reduction has been identified there is usually not much argument about the basic idea of installing a safety related system and often this is a safety instrumented system. Justification issues arise when the scale of investment has to be decided ... is it to be a cheap system with high running costs or a more expensive model that repays its cost in reduced operating and maintenance costs? Searching out the links between SIS costs and true running costs of the plant may be a tricky job.

In practice engineers and managers have to make choices on the type, quality and costs of the safety solutions available within the constraints imposed by the essential safety requirements.

- The type of solution may mean using mechanical devices such as relief valves and water spray system instead of sensors and control valves.
- The quality may mean choosing between a relay based SIS and a dual redundant PES.
- It may be that a safer but more expensive process is a better solution than a risky but efficient process protected by an SIL 3 SIS.

13.1.1 Justification issues

Making the right decision and justifying them should be a lot easier if the true cost of the various options can be set out with a fair degree of credibility. The issues in justification of SIS scope and cost are therefore:

- Failure modes of safety systems and their effects on the business
- Who is responsible for the justification
- What are the life cycle costs
- Finding the optimum solution

The issue of credibility and acceptability of the answers to a justification exercise are certainly outside of the scope of this workshop.

13.2 Impact of safety system failures

By now we should be familiar with failure modes but each type has a different potential for impacting the cost of the business. The modes are outlined in the next four figures

13.2.1 Mode 1: dangerous undetected failures of the SIS

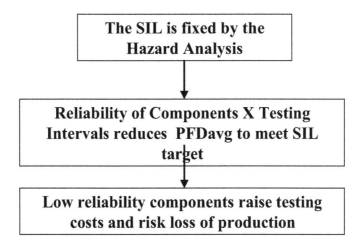

Figure 13.1
Impact of failure modes...1

This mode of failure is governed by the SIL requirements extracted from the hazard analysis. There will no dispute about meeting the target SIL. However, the frequency of testing can be raised to help a poor quality system meet the SIL. Hence operating costs may be forced up by buying a low specification SIS.

13.2.2   Mode 2: dangerous detected failures of the SIS

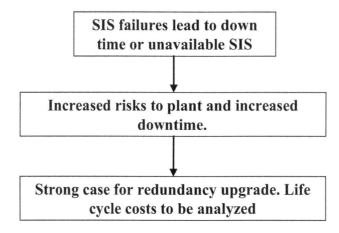

Figure 13.2
Impact of failure modes...2

This mode of failure (fail safe action or found by testing) results in interruptions to the production process and hence loss of income for the business. If repairs are attempted on-line the period during which the process is unprotected rises. Hence operating costs may be forced up by buying a low specification SIS. For this type of failure a high spurious trip rate is often a justification for upgrading an older SIS.

13.2.3   Mode 3: degraded mode of a redundant SIS

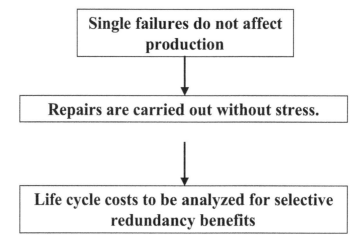

Figure 13.3
Impact of failure modes...3

This mode of failure involves a redundant system reducing to a single or 1002 mode of protection. The level of protection remains high and the shutdown rate is very low. Initial installation costs are likely to be higher but the life cycle cost may be lower.

13.2.4   Mode 4: nuisance trip failures of the SIS

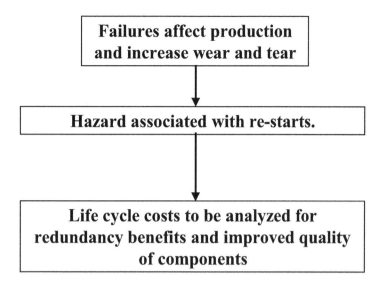

Figure 13.4
Impact of failure modes...4

Similar to mode 2 in effect but leaving no choices on production losses. All forms of shutdown mean a loss to the business. In addition there are potentially increased costs for wear and tear on the main plant equipment as crash shutdowns occur. There is often an increased risk of hazards due to the disturbances caused by an unscheduled trip followed by the risks of operation under hastily recovered start up conditions. Measures to reduce spurious or nuisance trips are therefore likely to show benefits for the life cycle cost.

13.3   Justification

13.3.1   Responsibilities

Whilst justifying an SIS on the grounds of meeting the safety requirements is done fairly readily by managers and engineers the justification for improved overall performance of the SIS is a technical responsibility for the Control or Electrical Engineer. Some of the benefits of operating with a better class of SIS equipment such as a PES-based logic solver are very difficult to quantify. However we have to make a case from some reasonable foundation and the best tool for supporting a case is the life cycle cost analysis.

13.3.2   Life cycle cost method

Life cycle costing presents the total cost of an installation in terms that a business man will understand and appreciate. The true cost of ownership of any plant item is measured by this method as an equivalent lump sum price in the present day money.

The next figure is tree structure diagram showing typical components of life cycle cost for a safety instrumented system. A more comprehensive list would have to show details of all project related costs. For annual costs the model has to includes items such as the service agreement and software licensing costs for the logic solver. A model with suitable headings soon prompts the user to fill in details of related cost items. This is not difficult to set up if you know your particular operation quite well. Our example here is very minimal.

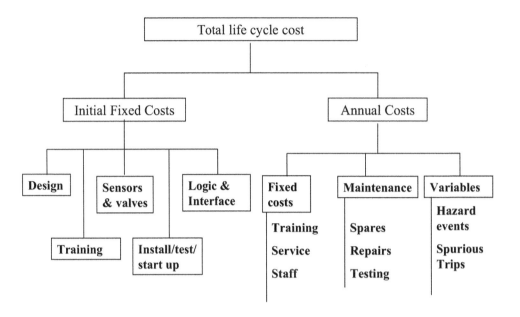

Figure 13.5
Cost breakdown structure for SIS

Use the structure diagram to develop and expand the headings for costs. The difficult part is to put real numbers in the boxes!

In particular it may be contentious to try and include an annualized figure for the cost of hazardous events. This is not so difficult to justify in the case of asset loss applications. For example the cost of failure of a turbo-compressor protection system can be measured as the price of a new rotor plus the cost of lost production.

The next step will be to set up a costing spreadsheet for the standard and agreed version of life cycle costs so that various scenarios can be tested for cost. The next figure is a simplified model for the spreadsheet.

Factor	Initial or Fixed Costs	Material	Labour	Totals
	Design			
	Training on logic solver			
	Sensors and valves			
	Logic and interface incl. config.			
	Install/test. Start up and validation			
	Fixed Cost Sub -total			
	Annual Costs			
	Fixed items: (staff, training, building)			
	Maintenance/spares/repairs			
	Service Agreements/sw licences			
	Testing			
	Hazardous events (D xPFD)			
	Spurious trips (λs)			
	Annual Costs Sub -total			
	Present value for annual costs over 20 yrs.			
Total Life Cycle Costs				

Figure 13.6
Example of a typical life cycle cost table

Note the method of representing the accumulated value of annual maintenance costs. The value is expressed as the equivalent of an investment cost made in the present. An assumed life for the SIS has to be used.

13.3.3   Costing example

The evaluation method is best illustrated by taking an example of a typical application in a process plant. Here we have to imagine there is a process plant with say 8 separate safety functions all served by a common PES logic solver with an interface to the plant DCS.

The next figure shows a reliability block diagram for a single loop function in the plant that shuts down the operation if the conditions become hazardous. The reliability data used is arbitrary but may be realistic for installed performance as opposed to manufacturers' product data.

Reliability model: case 1
Case 1 employs a dual 1oo2 sensor pair as inputs to a single channel logic solver with diagnostics, 1oo1 D. A dual redundant pair of valves is used to shutoff feed to the process. The safety requirements specification for this function calls for SIL 2 integrity to protect against the hazardous condition that could arise as much as once per year (i.e D =1/yr).

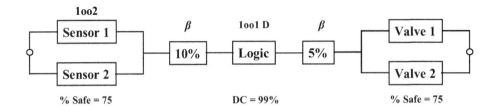

λ	1×10^{-5}	5×10^{-5}	1×10^{-5}
MTBF$_{SP}$	5 yrs.	2 yrs.	5 yrs.
PFDavg.	1.1×10^{-3}	2.7×10^{-4}	5.8×10^{-4}

Overall PFD = 2×10^{-3} Overall SIL = 2
Spurious Trip Rate = 1/5 + 1/2 + 1/5 = 0.9/yr.

Figure 13.7
Reliability model: case 1

Reliability analysis results as shown in Figure 13.7 indicate that the SIL 2 target can be met with the assistance of proof testing at 2 times per year. The nuisance trip rate is predicted at 1.25 times per year based on the sum of the spurious trip rates found for the 3 elements of the loop.

The issue here is: Can we find a viable way to reduce the production losses due to spurious tripping without compromising safety? We need to perform a benefits analysis on any proposed upgrade to get a feel for the options and ultimately to be able to justify the upgrade.

Reliability model: case 2

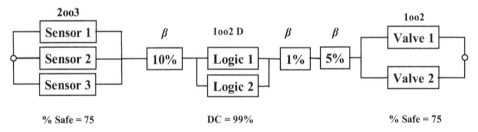

λ	1×10^{-5}	5×10^{-5}	1×10^{-5}
MTBF$_{SP}$	40 yrs.	100 yrs.	5 yrs.
PFDavg.	1.2×10^{-3}	1×10^{-4}	5.8×10^{-4}

Overall PFD = 1.8×10^{3} Overall SIL = 2
Spurious Trip Rate = 1/40 + 1/100 + 1/5 = 0.22/yr.

Figure 13.8
Reliability model: case 2

We show here one possible upgrade scenario and call it case 2. The sensor pair have been upgraded to 2oo3 voting to reduce their spurious trip rate. The logic solver is to be upgraded to 1oo2D for the same reason, but keep in mind this will benefit the spurious trip rate for all 8 safety functions. In this example the plant costing models share the basic PES logic solver costs equally across each function but the cost of upgrading to 1oo2D will be attributed to the single function. In practice the full cost benefit analysis would have to generate a cost sheet for each function.

13.3.4   PFD Comparisons

Case 1 meets SIL 2 with proof testing 2 times per year
PFDavg is .002 and hazard rate is (D×PFD) = 1×0.002/yr

Case 2 still meets SIL 2 with proof testing 2 times per year
PFDavg is .0018 and the hazard rate remains unchanged.

In this example the PFDavg is virtually unchanged by the new design and the SIS easily meets SIL 2. Note that we can't save costs on testing because the sensor group PFD is not changed very much by going to 2oo3 voting.

13.3.5   Nuisance trip comparisons

Case 1 nuisance trip rate = 0.9 per year
Production Cost = £ 30 000 × 0.9 = £27 000 per year

Case 2 nuisance trip rate = 0.22 per year
Production Cost = £ 30 000 × 0.22 = £6 600 per year

Additional costs for sensors and PES upgrade costs attributed to this function only. Benefits to other functions excluded.

Case 2 delivers a substantially improved nuisance trip rate mainly through benefits gained in MTBFsp figures for sensors and PES. The redundant valve pair are now the main contributors to spurious trips but moving these to a 2oo3 design presents an increase in complexity that may outweigh the benefits.

13.3.6   Cost comparisons

Now we fill in some cost data on the tables to see how the two cases compare. Again the data is arbitrary and does not imply any standard value of costs.

Life Cycle Costs Table for Case 1

Factor	Initial or Fixed Costs	Material	Labour	Totals
1	Design		10 000	
0.125	Training on logic solver		5 000	
1	Sensors and valves	21 000		
0.125	Logic and interface incl.config.	10 000	3 000	
0.125	Install/test. Start up and validation	4 000	6 000	
	Fixed Cost Subtotal	35 000	24 000	59 000
	Annual Costs			
0.125	Fixed items: (staff, training, building)		5 000	
1	Maintenance/spares/repairs	2 000		
0.125	Service Agreements/sw licences	3 000		
1	Testing		2 000	
0.002	Hazardous events (DxPFD) @ £ 10^6/event	1 000	1 000	
0.9	Spurious trips (λs) @ £ 3 x 10^3/event	27 000		
	Annual Costs Sub-total	33 000	8 000	
	Present value for annual costs over 20 yrs.	396 000	96 000	492 000
Total Life Cycle Costs				551 000

Figure 13.9
Life cycle costs table for case 1

Case 1 indicates a total investment of £ 551 000 including for the possible cost of then only spurious trips at rate of 0.9/yr

Life Cycle Costs Table for Case 2

Factor	Initial or Fixed Costs	Material	Labour	Totals
1	Design		10 000	
0.125	Training on logic solver		5 000	
1	Sensors and valves	24 000		
0.125	Logic and interface incl.config.	30 000	3 000	
0.125	Install/test. Start up and validation	6 000	8 000	
	Fixed Cost Sub-total	60 000	26 000	86 000
	Annual Costs			
0.125	Fixed items: (staff, training, building)		5 000	
1	Maintenance/spares/repairs	3 000		
0.125	Service Agreements/sw licences	4 000		
1	Testing		3 000	
0.002	Hazardous events (DxPFD) @ £ 10^6/event	1 000	1 000	
0.22	Spurious trips (λs) @ £ 3 x 10^3/event	6 600		
	Annual Costs Sub-total	14 600	9 000	
	Present value for annual costs over 20 yrs.	175 200	108 000	283 200
Total Life Cycle Costs				369 200

Figure 13.10
Life cycle costs table for case 2

Case 2 indicates a total investment of £ 369200 including for the possible cost of then only spurious trips at a rate of .22/yr.

The saving in annual cost of spurious trips offsets the increased capital cost (£ 27000) of the upgraded design by an improvement of £ 181800.

In practice we would have to extend this study to all safety functions sharing the logic solver. It may be that several of the other functions do not have much impact on spurious trips and in this case the loading factors shown in the table would have to be weighted for the critical items.

Clearly this is a simplified case but it indicates the approach to justification. A well verified spreadsheet model would of course enable many alternative designs to be evaluated. If the reliability analysis models can be supported by a software package such as the one we used earlier in the workshop the process can be made reasonably efficient.

The benefits of life cycle cost models can be seen in the improved perceptions of what each safety function is really doing for the business. The problem of credibility remains in the area of predicting losses from a probability based model. After all, 'it may never happen'.

13.3.7 Conclusion

The combination of reliability modeling and life cycle cost analysis can produce very useful data for use in the decision and justification tasks. The issue of credibility has to be taken into account, particularly when the savings are claimed for items dependent on probability analysis. What is clear is that many issues of design selection and operating philosophy are well supported by maintaining good cost models for the SIS.

Appendix A

Practical exercises

Exercise 1 – Calculating risk reduction factor and SIL

This practical exercise supports Chapter 2 with two basic exercises in calculating the risk reduction factor and SIL required from a safety function in a hazardous situation.

Subject	Calculation of RRF and SIL	
Objective	To ensure that the reader recognizes the basic relationships between risk reduction and safety integrity levels. To provide experience in the use of a risk classification table	
Relevance	Chapter 2. The quantitative method for determining target safety integrity levels	
Task detail	1) Use the starting information given below and refer to Table 2.5 in Chapter 2 to classify the given risk and its frequency.	
	2) Using this table, decide the maximum tolerable risk frequency to reduce the risk to class 3 (considered to be acceptable)	
	3) Calculate the target risk reduction factor, PFDavg values and safety availability required from the proposed safety instrumented system to achieve the tolerable risk frequency	
	4) State the target safety integrity level required from the SIS by reference to the SIL tables in Chapter 2.	
	5) Repeat the above for the 2nd example	

Starting information

	Potential Hazard/Risk	**Proposed Solution**
Example 1	A chlorine electrolyzer plant presents a major leak hazard due to loss of pressure control. The estimated frequency of occurrence is once per 10 years. The estimated consequence without any protective measures is that the operating team of 3 people will be likely to suffer serious injury or they may be killed. A school in the neighborhood may experience toxic fumes leading to injuries and a public outcry	A safety instrumented system will monitor pressure limits and will trip out the electrolyzer operation before the leak condition can arise.

Example 2	An industrial boiler is likely to burst a tube if the water level in its drum falls too low. If a tube bursts there is a risk of serious injury to the field operator who is often close to the plant. The boiler drum level is maintained by an automatic level control loop.
	
	The reasons for the low level could be
	  a) Failure of the level control system instrumentation or
	  b) Failure of the feedwater supply through loss of level in the feed tank or through failure of the feedwater pumps.
	
	The predicted failure rates are:
	Level Control: 1 dangerous failure per 5 years
	Feedwater supply failure: 4 times per year
		1 Install a pressure alarm in the BPCS on the feed-water supply alerting the operator who will have time to start another pump or shut down the boiler. The alarm combined with the operator's response is expected to work 4 times out 5.
		
		2 Additionally, install a drum low level sensor and trip system (SIS) to shutdown the boiler if the drum water level falls to the danger point.

Specimen answers to Exercise 1

Ex 1:

1. Event has more than 1 death: This puts it into the 'Catastrophic' column in the risk classification table.

2. Hazard demand rate 1/10yrs. So unprotected risk frequency, Fnp = 0.1/yr

3. Now determine the tolerable risk frequency:
       Class 3 risk requires < 1event per 5000 yrs . So Ft = .0002/yr.
    These frequencies can be shown on the IEC risk reduction model.

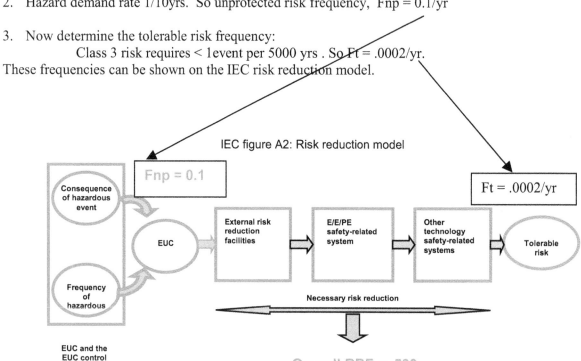

The above general risk reduction model simplifies to:

*   Risk reduction factor: RRF = Fnp/Ft = .1/.0002 = 500
*   The PFDavg required is 1/RRF = 1/500 = 2×10^{-3}
*   Safety availability = (RRF–1/RRF) = 499/500 = 0.998 or 99.8%.

4. The IEC or ISA table shows the required PFDavg is in the range 10^{-2} to 10^{-3} and therefore: **the required SIL is 2**

Exercise 2 – Hazard study

This practical follows from a study of the basics of hazard studies in Chapter 3. Its objective is to demonstrate a typical hazard study.

Subject	Application of hazard study 2 and 3 methods to a raw-gas holder	
Objective	To identify potentially hazardous events in a proposed design and to propose safety measures to reduce risk.	
Relevance	The study forms part of a typical safety life cycle activity	
Task detail	This exercise is described for a 'study team' of two or more persons. However it can also done by an individual. 1) Study the following 'starting information' which provides a description of a proposed gas holder system to be installed between the gas producer stage and the gas compression stage of an ammonia plant. (Typical hazard study 1 output.) 2) Use the hazard study 2 guidelines in Chapter 3 to identify potential risks. The study team should apply the guide diagram 1 to identify possible hazards and basic causes. Use guide diagram 2 to identify possible consequences. 3) The team then considers the possible sequence of events that might cause each of the hazardous events it has identified. To assist in this stage the team should make use of the hazard study 3 method outlined in Chapter 2. Even though this is a level-2 study the deviation prompting tools of hazard 3 are useful in this type of study. Use this method to examine the main gas flow route through the gas holder and consider deviations from normal conditions such as high flow or impurities. 4) The hazards, causes and consequences are to be recorded in the reporting form. 5) The team should then discuss measures to reduce the likelihood and/or the consequences of the events. These measures are to be recorded against each event in the reporting form. Follow up actions should be listed in the last column. 6) Finally the team should try marking on to the diagram any suggested SIS functions with sensors and actuators.	

Starting information

IEC safety life cycle phases 1 and 2 or process hazard study 1 provides the following information.

IEC ref	Requirement	Information
7.2.2.1	EUC and required control functions	See diagram 1 and description of process on page 3
7.2.2.1	Physical environment	Inside chemical complex but 100 meters from a canteen and recreational area for 500 persons used daily.
7.2.2.2	Likely sources of hazard	Large volumes of raw gas that is not flammable until mixed with oxygen. Oxygen is used in the upstream gasification stage, hence there is a possible risk of producing flammable or explosive gas mixtures. Potential for internal explosions or fires, external explosions or toxic gas releases.
7.2.2.3	Hazard info	Raw gas consists of 25% H_2, 60% CO, 10% CO_2, 5% other gases inc H_2S (smelly). Raw gas is flammable in range 7.3% to 74.5% in air. Toxic component is CO with a short term exposure limit of 400 ppm and immediate escape level of 1500 ppm.
7.2.2.4	Current safety regulations	OHS limits for toxic exposure.
7.2.2.5	Interactions with other EUCs	Gas holder volume (level) rises and falls in response to mismatch between upstream and downstream plant rates. Gas holder contents must be isolated if either upstream gasification stage or downstream processing stage is shutdown.
7.3.2.1	Physical equipment in the scope of the hazard and risk analysis	See diagram 2 and description on sheet 4
7.3.2.2	External events to be taken into account	1) Tripping of 1 to 6 gasifiers 2) Tripping of 1 or 2 gas blowers 3) Tripping of downstream plant 4) Tripping of 1 or 2 compressors 5) Flame out of gasifier or oxygen leak into gasifier output.
7.3.2.3	Subsystems associated with hazards	Cooling water supply. Raw gas flare system.

Information to be developed by the hazard study team

In this practical the results required from the hazard study team are **shown in bold type** in the following table which shows their relevance to the safety life cycle.

IEC safety life cycle phases 2 and 3, require the following information.

IEC ref	Requirement	How the information is obtained
7.3.2.4	**Type of accident initiating events that need to be considered**	**Determined by hazard study 2 method**
7.4.2.1	Hazard and risk analysis to be done	Follows after the risks have been identified
7.4.2.2	Consideration to elimination of the hazards	Part of hazard study, refers back to designers
7.4.2.1	**The hazards and hazardous events of the EUC and EUC control system (in all modes of operation) to be determined**	**Determined by hazard study 2 method. In this practical ignore start-up modes.**
7.4.2.4	**The event sequences leading to the hazardous events**	**Determined by hazard study 2 method**
7.4.2.5	**Likelihood of the hazardous event**	**Hazard study 2 to make an assessment.** Further studies assisted by fault tree analysis
7.4.2.6	**Potential Consequences**	**Preliminary from the hazard study 2.** Further details as part of risk analysis.
7.4.2.7	EUC Risk for each hazardous event	Determined by hazard analysis

Suggest risk reduction measures for any identified hazards.

(In preparation for the specification of overall safety requirements: IEC safety life cycle phase 4. The hazard study 2 is often used to propose the safety functions that will be incorporated into phase 4 after analysis.)

Raw gas holder: Scope and description of process plant (equipment under control).

See diagram 1.

A coal based ammonia plant has been designed with two raw gas production streams, each stream consisting of 3 gasifiers supplying a large gas blower. The gas from the blowers has to be cleaned by passing it through an electrostatic precipitator section before it is supplied to a pair of gas compressors. The compressors deliver the gas into a 'rectisol' stage in which the hydrogen is separated from the carbon monoxide and dioxide components.

The raw gas holder provides buffer storage of gas between the gas production and gas compression/rectisol stage to allow for short term mismatches between the rate of gas production and rate of compression. This is essential for practical start up of the plant and for coping with changes in the rates of gas production and gas compression.

The gas holder provides a variable capacity storage of 1000 to 10 000 cubic meters (m^3) of raw gas at a virtually constant pressure of 35 mbar gauge. (i.e. above atmospheric pressure). This is achieved by allowing the top section of the gas holder to rise and fall as the net flow into the gasifier varies from positive to negative.

The rate of gas production is 20 000 m^3/hr per gasifier, hence with all 6 gasifiers running the gas production rate will be 120 000 m^3/hr. The rate of the downstream gas compressor stage is to be controlled to match the inflow of gas so that the level of the gas holder is stabilized at typically 50%. The EUC control system comprises an automatic level control loop operating on the input control valves of the compressors supported by manual controls that allow the operator to decide if one or two compressors are to be run to balance the load.

One compressor can handle 1 to 3 gasifiers and two compressors can handle 3 to 6 gasifiers. Hence if 1 gasifier suddenly trips out the compression rate must be reduced by 16.6% or 20 000 m^3/hr to rebalance the flows. The gasholder capacity at 50% level provides a reserve of 5000 m^3, equivalent to 15 minutes of mismatched operation before the capacity falls to zero. In another scenario if 1 gas blower trips the compression rate may have to be reduced by 50% and the re-balancing time available may fall to 5 minutes.

Physical equipment in the scope of the hazard and risk analysis

See diagram 2: A simplified version of the equipment is used for this practical

The EUC scope for hazard study covers the gas holder and the precipitators stage starting at the discharge point from the gas blowers and ending at the inlet point to the gas compressors. The equipment includes a flare stack used to divert and burn off raw gas when starting up the gasifiers or in emergency conditions.

The gas holder consists of a 30 meter diameter domed top section floating above a water filled bottom section. As gas flows into the top section it will rise between spiral guide rails in response to the small rise in pressure.

The internal gas space is sealed by water which is maintained at constant level in the bottom section and which is pushed up the sides of the bottom tank by the pressure in the dome. The sides are open to atmosphere and hence provide an ultimate pressure relief for the gas holder.

A drainage system is provided at the bottom of the inlet and outlet pipe u-sections or lutes to continuously remove condensate deposited by the incoming gas as it cools. Hence a pressure sealing water trap or seal pot is provided. When the plant is shutdown these lutes are used as isolators to trap gas in the holder by flooding the lutes from a water tank.

The EUC control system comprises:
- The level control loop described above
- Remote manual controls to adjust the compressors
- High and low level alarms on the gasholder level taken from the level controller
- Remote manual controls for opening and closing the flare
- Remote manual controls for motorized valves used to isolate the gas holder by flooding the lutes.

The precipitators are large volume rigid chambers with electrostatic plates, which attract the charged dust particles in the gas and hence clean the gas stream. The precipitator plates are likely to produce sparks.

Reporting forms for exercise 2

Haz 2 Study Reporting Form	Exercise 2: Raw gas holder hazard study		Drg No: RGH Diagrams 1 and 2	Rev No: 1
Team Members:			Date	Sheet No of Meeting No.
Hazardous Event or Situation	**Caused by / sequence of events**	**Consequences** Immediate/ Ultimate	**Estimated Likelihood. / Suggested measures to reduce likelihood**	**Emergency Measures** (reduce consequences)	**Action Required**
External Fire	Ignition of gas escapes from RGH due to: 1: Loss of level control, • Due to plant disturbances exceeding the range of compressors • High level stops reached in RGH leads to gas pressure rise, leads to gas escape via lutes. Gas ignited by sources of ignition Or: 2. Due to loss of water in lutes due to evap or leakage. Gas escaping from lute is ignited	Limited plant damage and loss of production	**Probable** Loss of level control is probable. Ignition is possible if sources of ignition are not prevented. **Reduce likelihood by:** 1 Prevent lutes running dry by continuous feed of water and low flow alarm. 2 Protection against high level stops being reached: Trip to divert gas feed to flare on approach to high level in RGH 3 Remote locating of flare. 4 Haz area classification	1 Fire fighting equipment 2 Personnel prevented from access to lute areas.	1 Haz analysis study 2 Classify area 3 Request design of highly reliable water supply to lutes. 4 Manual monitoring or alarms on lute levels. 5 Low flow alarm on water to lutes.
Internal fire	Flammable mixture formed by residual air at start up or breakthrough of oxyxgen from gasifiers. Mixture ignites at precips		Moderate risk due to possible problems with gasifiers, High risk at start up. Reduce likelihood by: 1 Design gasifier controls to minimize chances of oxygen breakthrough, 2 Install oxygen detectors at Inlet to RGH. Trip the in-feed to flare on detection of oxygen. 3 Trip compressors as soon as trip to flare confirmed. Then	Consider heat sensors and alarms at precips. Shut off gas to precips and trip gasifiers	Hazard analysis required. Assessment of Risk and RRF. Specify SIS for protection against Oxygen. Note this is a very expensive trip event and spurious tripping is to be avoided.

			isolate RGH by opening flood valves to lutes.		
			4 Ensure full nitrogen purge before admitting raw gas. Test for oxygen at RGH exit during start up.		

Hazardous Event or Situation	Caused by / sequence of events	Consequences Immediate/ Ultimate	Estimated Likelihood. / Suggested measures to reduce likelihood	Emergency Measures (reduce consequences)	Action Required
Internal fire 2	Flammable mixture formed by entry of air to RGH. Sequence as follows: 1 One or more Gasifiers trip or a blower trips and compressor rates are not reduced to balance the flow. 2 Negative pressure occurs if RGH reaches low level stops and compressors continue to extract gas faster than inlet rate from gasifiers. 3 Air entry occurs if excessive negative pressure condition arises in RGH. 4 Mixture Ignites at precips.	Potential for explosion in RGH as below	Probable due to frequent tripping of gasifiers and the need for operator intervention to balance the compressors. Reduce likelihood by: 1 High Priority alarm on loss of gasifier or blower, urgent operator action to reduce rate of compressors 2 Tripping of both compressors if all gasifiers or blowers trip 3 Tripping of one compressor if low level approach occurs in RGH 4 Tripping of one compressor if one blower stops 5 Tripping of both compressors if low level limits are reached on RGH	As above	Hazard analysis required. Assessment of Risk and RRF. Specify SIS
Internal explosion	As above but fire burns back to gas holder where flammable mixture exists in large volume	Severe damage, rupture of gas holder, fragments Fatalities, injuries	Probable as for internal fire. Reduce likelihood by measures as above. Also • Design to prevent flashbacks from precips. • Isolation valves to be operable from control room. • Ensure emegency power always available to Isolation valves.	Design RGH to blow out upwards. I.e explosion panels in roof	Hazard analysis required Hazard analysis required. Assessment of Risk and RRF. Specify SIS

| Unconfined explosion | Flammable gas escapes and forms a vapor cloud in air. Reason Same as above for external fire | Loud noise, Missiles, blast damage, fatalities and injuries, to employees, not to public? (investigate), Large number of people close by. | Rare event due to light gas rising and diluting in air. Possibility cannot be ruled out. Protection as for external fire | Training of emergency services | Design review to improve equipment. Hazard study and risk assesment. Evaluate RRF Specify SIS |

Hazardous Event or Situation	Caused by / sequence of events	Consequences Immediate/ Ultimate	Estimated Likelihood. / Suggested measures to reduce likelihood	Emergency Measures (reduce consequences)	Action Required
Harmful exposure	1 Toxic gas escape from RGH. Events as for external fire but without ignition 2 Gas cloud blows across to adjacent plant area where high density of people exists 3 Or: Due to maintenance personnel entry to vessels and lutes	Acute effect on employees. Not on public. CO exposure from a large gas release eg at 2000m/1hr Large number of people close by likely to be injured employees. Ill health Bad publicity	More probable than external explosion. Large population of adjacent plant is at risk, 1 fencing to keep people further away 2 trip systems as for external fire	Area alarm system to alert staff to evacuate to safe areas or gas proof room.	Note that this event has great potential for harm. Hence the combined risks of gas leakage amount to a serious hazard. SIS must be designed to minimize the chances of an overpressure at the RGH.
Chronic exposure	Toxic gas: low level leakage: exposure mechanism: seals on valves maintenance	Chronic effect on employees/ill health	Probable 1 design for low leakage equipment, avoid valve sterns etc. 2 provide personnel with gas detectors,		
Pollution	Smells from sulphur in gas low level leakage: exposure mechanism: seals on valves		Probable Design for low leakage		

Conclusion for the safety requirements spec:

The following functional safety requirements have been identified:
High level approach requires an urgent alarm
High level limit requires a trip to flare
Low level approach requires an urgent alarm for operator to reduce compressor load
Low level limit requires immediate trip of compressors. Probably followed by trip of in feed to flare.
Oxyxgen in feed to RGH requires a trip to flare, trip of compressors to stop oxygen reaching the next stage and then requires isolation of the RGH to prevent backflow and mixing with the in feed.
The diagram overleaf has been marked up to show the suggested SIS protection loops.

The following notes may be of help.

The level sensors used to initiate trips are separate from the process controller to avoid common cause failures.

- Continuous transmitters are used so that their operating conditions can be diagnosed at all times.
- Redundancy is suggested despite the cost due to the difficulty of the measurement.
- Two different types of level measurement might be a good idea to include diversity and avoid common cause errors. A reliability analysis should be carried out to find the best configuration.
- In the original implementation of this function the designers decided to use switched level detectors and install them as 2oo3 voting units. The benefits of continuous measurement were lost but the practicality of installing a reliable switching device outweighed this benefit. Many level switch sensors include good diagnostics which improve their fail-to-safety performance and reduce their fail-to-danger rate.

There is no requirement for additional actuators for example at the flare because the cause of the problem was not expected to be anything to do with the existing actuators, hence no risk of a common cause problem.

The oxygen detectors are arranged in a 2oo3 voting system since the risk of spurious trips from an analyzer device is high and their overall reliability is not so good. The cost of a trip is very high so the returns for having a high safety availability combined with a low spurious trip rate are attractive.

Exercise 3 – Fault trees

This practical exercise is to construct a fault tree diagram using the basic principles introduced in Chapter 3. It uses an example of a simple reactor with automatically controlled feeds that has the potential to cause a serious risk to plant personnel. Once the basic fault tree has been drawn, the model is to be adjusted to incorporate a safety-instrumented system and to demonstrate the resulting risk reduction.

Subject	Application of a fault tree analysis to a chemical reactor	
Objective	To model the relationship between possible faults and hazards in a proposed design and to show the effects of risk reduction measures.	
Relevance	The study forms part of the detailed risk analysis that follows from a hazard study. It also shows how the fault tree model can be used to visualize the action of an SIS and compute its effect on the risk frequency.	
Task detail	6) Draw a fault tree for the example of a reactor hazard given in the starting information below. 7) Calculate the explosion rate for EUC without protection. 8) Add into the fault tree a simple trip protection against high fuel flow or low oxidant flow. 9) Calculate the new explosion rate. In constructing the fault tree and applying its rules it may help to refer to the guidance in appendix 3.1	

Starting information

Ref	Requirement	Information
Task 1	Fault tree for reactor explosion based on 'explosion' as the top event	Diagram 1 shows a reactor with a continuous feed of fuel and oxidant. An explosion can occur inside the reactor if the mixture becomes explosive and a source of ignition is found. In this case we might suppose the source is a hot catalyst inside the reactor presenting a 75% probability of igniting the explosive gas mixture when it occurs. The mixture can become explosive if the fuel flow becomes too high relative to the oxidant flow. Include for the possibility of sudden loss of feed in either stream. The failure modes of the control loops are to be shown as events based on faults in the sensors and the controllers. Assume controllers are kept on Auto.
Task 2	Calculate the explosion rate	Assume all instrument failure rates are 0.1/yr Assume the feeds fail at 0.2/yr
Task 3	Model the effect of fuel flow trips	On the fault tree, add the logic for a high fuel/oxidant flow ratio trip to shutdown the fuel supply. This arrangement is shown in diagram 2
Task 4	Calculate the new explosion rate.	Assume a PFD of 0.02 for the trip. Do not decompose the elements of the trip at this stage

Diagram 1 : Reactor for fault tree practical

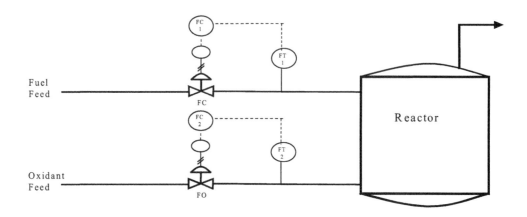

Diagram 2: Reactor with Flow Ratio Trip Protection

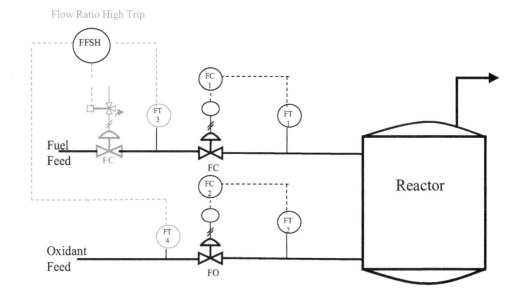

Answers to Exercise 3 – Fault trees

The top event is the explosion of a dangerous mixture. The probability of ignition is set at 0.75 as input to the AND gate. Two potential conditions for a dangerous mixture exist. The reasons for each type of condition are then developed downwards to find instrument loop defects that could be the causes. Figure 1 shows the resulting fault tree without any safety system protection.

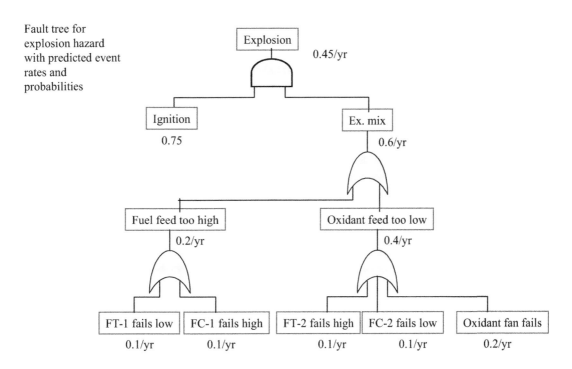

Fault tree for explosion hazard with predicted event rates and probabilities

Figure 1
Fault tree for unprotected hazard

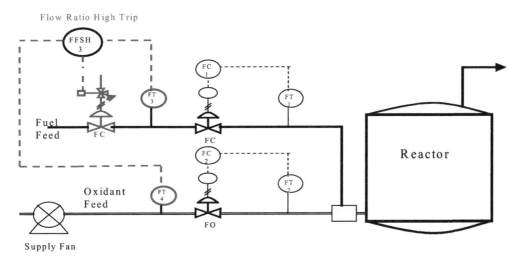

Figure 2
SIS function added to the process

Whilst it is clear that this process scenario is unlikely to be as rudimentary as this example we are going to assume that an SIS is to be fitted to monitor the fuel/oxidant ration as an independent measurement and that it will have a trip valve to close off the fuel. Figure 2 shows the SIS applied to the process.

Then the fault tree has to be modified to insert the possibility of a failed SIS AND the dangerous flow conditions before the dangerous mixture can occur. This is shown in Figure 3.

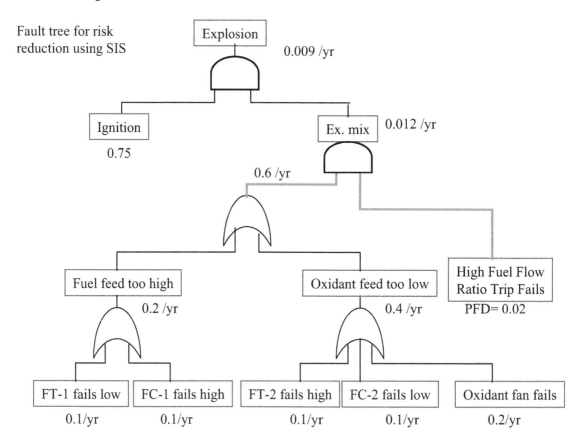

Fault tree for risk reduction using SIS

Figure 3
Fault tree with SIS added

The failure rates applied to the fault tree logic are processed by the rules of the AND and OR gates and predict the new explosion rate at 0.009/yr.

Exercise 4 – Risk reduction modeling

An example of a risk situation is used in this exercise. We are asked to use 'layers of protection analysis to arrive at a risk reduction model for the situation. The quantitative analysis method is then used to define the safety integrity level (SIL) required for the safety instrumented system. This model can also be used to check the practical application of qualitative methods for determining SILs.

Subject	Risk reduction model for a polymer autoclave reaction runaway
Objective	To define the layers of protection proposed for a polymer autoclave to reduce the risk of exposing site personnel to toxic vapors. To draw a risk reduction model incorporating the layers of protection and use the model to decide the required SIL for the safety instrumented system.
Relevance	Practices the use of IEC risk reduction modeling techniques. Illustrates the need to design an SIS in the context of overall protection as required by the safety life cycle.

Task detail	1) Identify the layers of protection provided in the given process information
	2) Draw a simple block diagram to show the layers
	3) Draw a risk reduction model and mark in the parameters for this application derived from the given data
	4) The parameters must include:
	   – Tolerable frequency (Ft) for a tolerable risk target of 0.1 persons/year exposed to toxic gases for a site with 50 people
	   – Unprotected risk frequency (Fnp)
	   – Overall risk reduction required
	   – PFD figures for each risk reduction stage (Note that the SIS and chemical inhibitor can be combined into one stage)
	   – Risk frequencies after each layer of protection has been added in
	5) Show the required PFD for the SIS as undetermined value = Y
	6) Calculate Y to yield the figure for Fp drawn on the model
	7) Check the applicable Demand Mode and state the SIL required for the SIS

Starting information

Ref	Requirement	Information
Task 1	Layers of protection	Diagram 1 shows an autoclave used to carry out polymerization batch operations typically as used in the manufacture of PVC. The autoclave is filled with monomer in a water mix and a catalyst is added. Heating is applied by the TIC loop until reaction starts to occur and the process becomes exothermic. The TIC loop applies cooling and stabilizes the temperature of the contents over a period of several hours as the contents convert to polymer. The risk of a runaway reaction occurs due to failures in the cooling system or in the control system or in chemical composition. When this occurs a high temperature alarm is used to warn the operator that the reaction is going too fast. Operator responds with manual override actions on the cooling system to try to recover control. Diagram 2 illustrates the stages of runaway that may occur if the alarm and operator action fail to solve the problem.

Ref	Requirement	Information
		If the reaction still gets away the safety instrumented system will automatically release a chemical dosage (or bomb) from a pressurized chamber. This is intended to kill the reaction and stop any further pressure rise. The pressure will fall as the temperature falls due to the loss of exothermic reaction. The chemical dosage has an estimated probability of failure on demand of 0.01. The PFDavg of the SIS is to be decided later.
		If the reaction does not respond to the dosage or if the trip mechanism fails the pressure will continue to rise. There is an extra-high pressure alarm which is set to warn of an impending release from the bursting disk. Site staff are trained to go to safe areas when this alarm sounds all around the plant. The result of this 'alarm and evac' procedure is that on average (including allowance for alarm failures) only 1 person in 10 is placed at risk when a bursting disk release occurs.
Task 2	Calculate the tolerable event frequency for the site.	50 people may be exposed to gas if the bursting disk blows. The company has decided that the toxicity risk dictates no more than 1 case of exposure should occur per 10 years.
	Calculate the unprotected risk frequency.	The risk of a runaway reaction occurs due to failures in the cooling system or in the control system or in chemical composition. This event occurs on average 2.5 times per year.
Task 3	Calculate risk reduction factors or PFDavgs for each risk reduction layer. Mark on the model.	The first alarm and the operator's response are estimated to solve the problem 3 times out of 4 events.
		The chemical dosage has an estimated probability of failure on demand of 0.01. The PFDavg of the SIS is to be decided later so is given the parameter Y.
		The result of the 'alarm and evac' procedure is that on average (including allowance for alarm failures) only 1 person in 10 is placed at risk when a bursting disk release occurs.
Task 4	Calculate Y, the PFDavg for the SIS. Look up the SIL value in the IEC table.	Y is the figure needed to make the model deliver the tolerable risk frequency. Check the demand rate on the SIS given by the model to decide on high or low demand mode operation

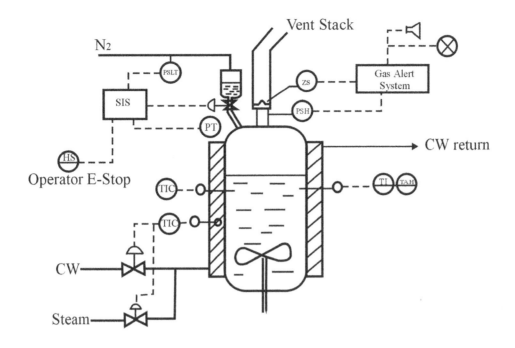

N₂

Vent Stack

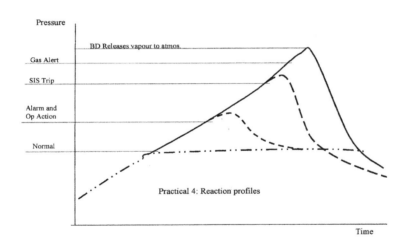

Diagram 1: Polymer autoclave with temperature controlled heating and cooling. Temp alarm is used to alert operator to loss of temperature control. SIS is used to stop runaway reactions. Gas alert is used to warn of impending release from bursting disk.

Diagram 2: Pressure profiles

Chart above shows the pressure rise in the autoclave as reaction rate increases. The initial heating phase changes to cooling as the reaction becomes exothermic. If the reaction rate is too high for the cooling system or if a controller fault develops the pressure and temperature will rise rapidly leading to a bursting disk release of vapours

Answer to Exercise 4

Task 1

The layers of protection are drawn thus

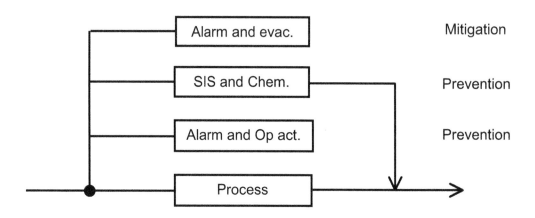

Task 2

We use a method of modeling the SIS and frequency requirements that shows actual event frequencies and allows for a group of 50 people to be at risk. The first two layers of protection contribute to reducing the event frequency for the risk of harming 50 people. The alarm and evacuation layer of protection contributes to reducing the consequence by protecting 90% of the people.

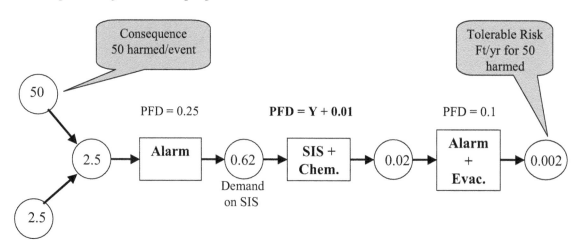

Unprotected risk
Fnp = 2.5/yr for harming
50 people

Tolerable risk is quoted as 1 person injured per 10 years. So Ft is 0.1/yr per person. For a group of 50 people exposed to risk this means a tolerable rate of 0.1/50 events per year = 0.002/yr.

The alarm and evacuation system effectively reduces to 5 the number of people who could be harmed. Hence the tolerable event rate for the gas release event when the alarm/evacuation layer is present is $0.1/5 = 0.02$/yr.

The temperature alarm response has a PFD of 0.25. This reduces the demand rate on the SIS and chemical bomb system from 2.5 to $(0.25 \times 2.5) = 0.625$ events/yr.

Then the combined PFD of the SIS and the chemical bomb is $(Y + 0.01)$
$Y + 0.01 = 0.02/0.625 = 0.032$
Hence $Y = 0.022$ i.e. the required PFD of the SIS $= 0.022$

The demand rate on the SIS is less than 1 per year so the mode is 'low demand mode' and the SIL = 1

The completed risk reduction model looks like this:

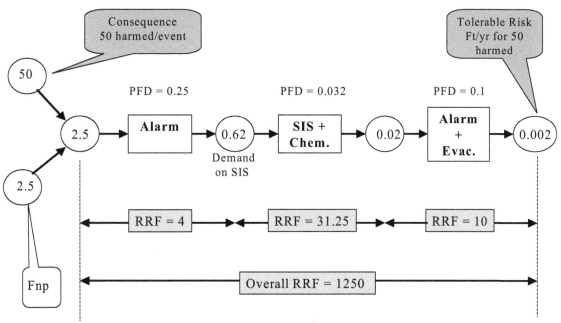

It is also possible to draw this model as a fault tree diagram. The equivalent fault tree looks like this:

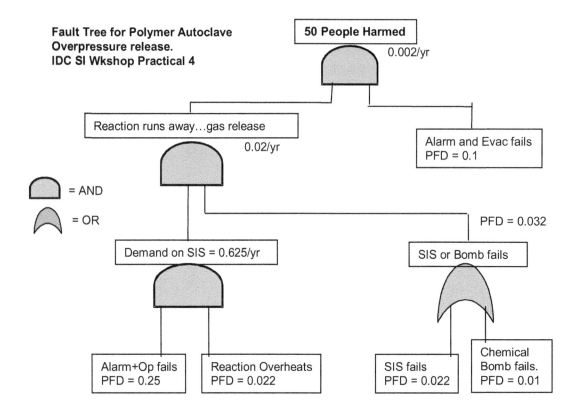

Fault Tree for Polymer Autoclave Overpressure release.
IDC SI Wkshop Practical 4

50 People Harmed — 0.002/yr

Reaction runs away...gas release — 0.02/yr

Alarm and Evac fails PFD = 0.1

= AND

= OR

Demand on SIS = 0.625/yr

PFD = 0.032

SIS or Bomb fails

Alarm+Op fails PFD = 0.25

Reaction Overheats PFD = 0.022

SIS fails PFD = 0.022

Chemical Bomb fails. PFD = 0.01

Exercise 5 – Reliability models

This practical exercise requires us to draw up analysis models for two configurations of safety instrumented systems and then calculate their spurious trip rates and PFDs. Comparison indicates the benefits of redundancy.

Subject	Reliability modeling and calculation.	
Objective	To find the best combination of instruments for a given safety requirement.	
Relevance	To demonstrate the quantitative method of evaluating the SIL for a proposed trip.	
Task detail	1) Calculate the spurious trip rate and the fail to danger rate for a single channel trip loop on an overheat protection example. 2) Model and calculate the same for a 2oo3 sensor arrangement.	

Starting information

Ref	Requirement	Information
Task 1	Calculate single channel failure rates for the high temperature trip example. Use the single channel analysis block diagram (diag. 2) for reference. Calculate the failure rates and PFDs for each stage. Sum them to give the overall single channel spurious trip rate and the PFD.	Diagram 1 shows a high temperature trip on a reactor. Three temperature transmitters are provided, each of which transmits to a trip amplifier device that acts as a high temperature trip device. In the first version of this design a relay system is used to operate a 2oo3 voting function leading to a single channel actuation. The trip actuation uses a solenoid valve and to vent the air cylinder on a valve that will drive open and release quench water into the reactor. A single channel version of the analysis model is shown in Diagram 2. The assumed fault rates are given in the table. Assume a single channel action uses 2 relays, 1 in the input stage and 1 in the logic solver. The first task is to calculate the spurious trip rate and the fail to danger rate (PFDavg) of a single channel version on the basis of proof testing every 6 months.
Task 2	Calculate the spurious trip rate and the fail-to-danger rate (PFD avg) of the 2 out of 3 channel voting version. First draw the analysis block diagram for this version.	Assume the 2oo3 relay stage uses 1 relay for the logic by using multiple contacts from the input stages. For the PFD and Spurious trip calcs assume the common cause factor is 10% and the proof test interval remains at 6 months.

Diagram 1: Reactor with high temperature trip using 2oo3 sensor architecture.

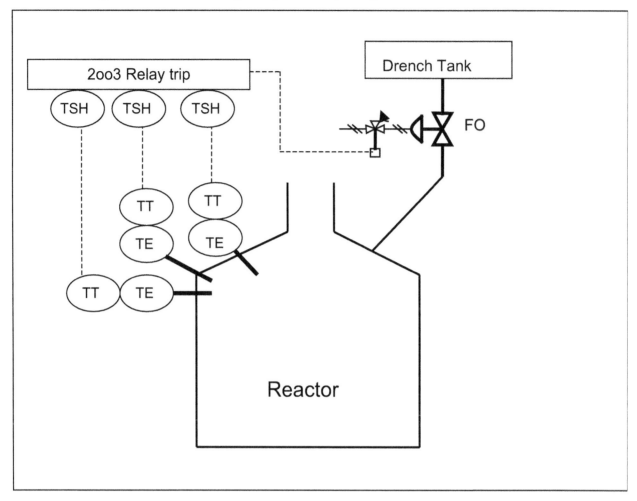

Table of fault rates for the devices in Diag 1

Channel Device	Fail-safe rate per year	Fail-danger rate per year
TE…element	1.5	0.20
TT…Transmitter	0.5	0.05
Cable/terminals	0.01	0.00
TSH….trip amplifier/switch	0.5	0.1
Relay (each)	0.05	0.002
Solenoid Valve	0.04	0.02
Trip Valve	0.4	0.1

Diagram 2: Single channel SIS arrangement. Reliability block diagram outline shown below.

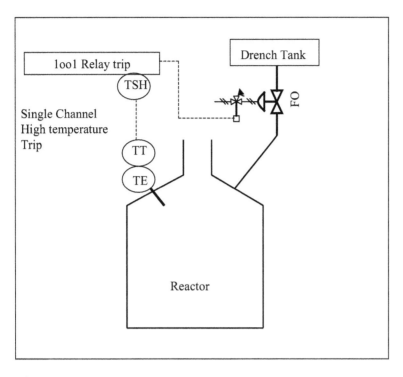

Hints:
Arrange fail-to-safe fault rates across the top of the model
Arrange fail-to-danger rates across the bottom of the model
Sum rates into sensor, logic and actuator groups

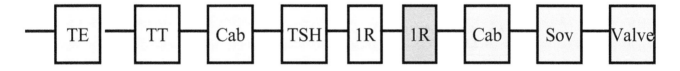

Answers to Exercise 5 – Reliability models

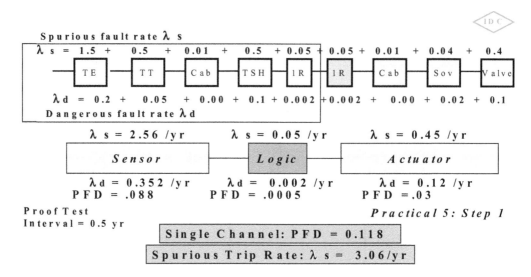

Figure 1
Reliability model for single loop

Fail-to-safe and fail-to-danger rates are assembled for all components acting in series and then they are grouped into the sensor, logic and actuator stages. The PFDs for each stage are shown and the system PFD is obtained by summing the series components.

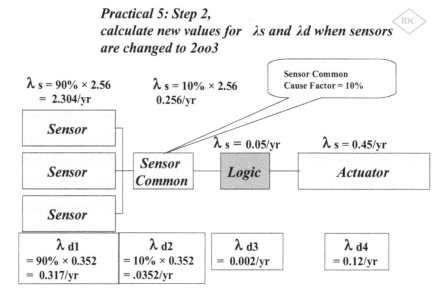

Figure 2
Reliability model for 2oo3 sensor stage

The PFD for the 2oo3 sensor stage is then calculated using the formula shown in Figure 3. The new PFD is added to the existing PFDs for the remaining stages of the SIS.

Practical 5: Step 3, calculate new PFD values

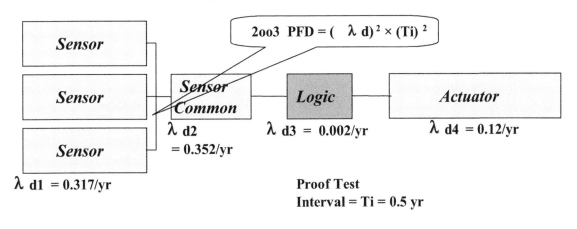

$$\text{2oo3 PFD} = (\lambda d)^2 \times (Ti)^2$$

λ d2 = 0.352/yr

λ d3 = 0.002/yr

λ d4 = 0.12/yr

λ d1 = 0.317/yr

Proof Test Interval = Ti = 0.5 yr

PFD = 0.025   PFD = .0088   PFD = .0005   PFD = .03

Overall PFD = 0.064

Figure 3
Reliability model with 2oo3 sensor stage showing calculation of PFDs

Practical 5: Step 4, New Spurious Trip Rate for 2oo3 section

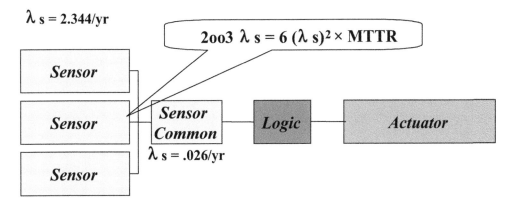

λ s = 2.344/yr

$$\text{2oo3 } \lambda s = 6 (\lambda s)^2 \times \text{MTTR}$$

λ s = .026/yr

Let MTTR = 24hrs = 24/8760 yrs = 0.0027yr

2oo3 λ s = 6 (2.344)2 × .0027 = 0.089

Spurious Trip Rate for 2oo3 section: λs = .089 + .026 = 0.115/yr

Figure 4
Reliability model with 2oo3 sensor stage showing calculation of spurious trip rates

Practical 5: Step 5, New Spurious Trip Rate
for overall loop

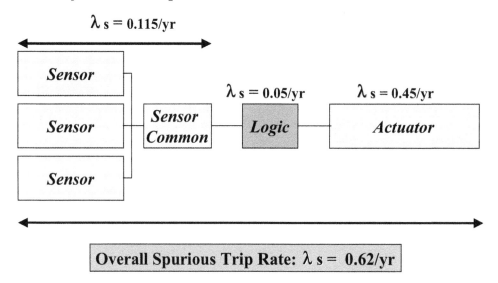

Figure 5

Reliability model with 2oo3 sensor stage showing calculation of overall spurious trip rate

Finally the information is assembled to show the comparative results for 1oo1 and 2oo3 designs.

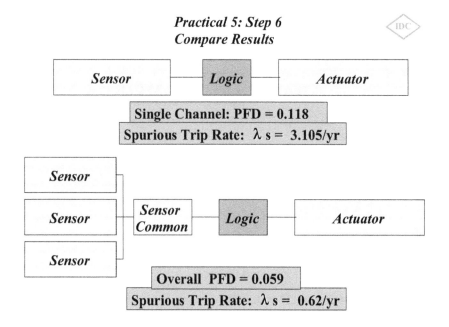

Figure 6

Reliability model with 1oo1 and 2oo3 sensors stage showing comparison of PFD and spurious trip rates

Exercise 6 – Evaluation of an SIS output stage

This is an exercise in evaluating the safety integrity of a proposed design for a trip system output stage where options have to be considered for a control valve to be used as a trip valve or as a redundant back up to a trip valve.

Subject	Architecture design for final element stage
Objective	To demonstrate a practical method for deciding on the use of a shared control and trip valve
Relevance	Chapter 7. Field Devices selection
Task detail	1) Option 1: Find the trip failure rate for an SIS designed as per Figure 1 where V1 is trip valve and V2 is a control valve with the trip command forcing it to close. Do this by constructing a fault tree analysis diagram. See guidelines below in the starting information. 2) Option 2: Deduce the failure rate if V1 is removed from the design. 3) Option 3: Deduce the failure rate if the SIS is connected only to trip valve V1 leaving V2 purely as a control valve. 4) Option 4: Deduce the failure rate if V2 is kept as a trip valve along with V1 whilst a new control valve is fitted for the basic control duty alone.

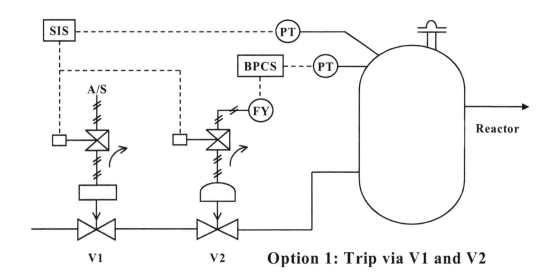

Option 1: Trip via V1 and V2

Starting information

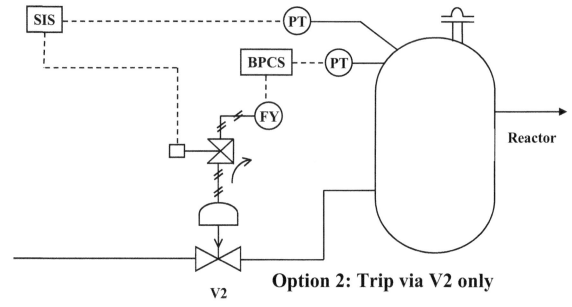

Option 2: Trip via V2 only

V2

The data given is as follows;

- λdd is the fail-to-danger rate of the combined solenoid valve, actuator and valve for V1 or V2. There are no undetectable failures.
- The proof test interval is Ti = 0.5 yr
- λdd for V1 = 0.2/yr, PFDavg for V1 operating alone = .05
- λdd for V2 = 0.2/yr, PFDavg for V2 operating alone = .05
- PFD for V1 and V2 in 1oo2 pair. PFDavg = .008 (based on common cause failure factor, β = 10% and manual proof test only, Ti = 0.5 yr)
- trip demand from process and control excluding failures of V2 is estimated to be 1/yr.
- The combined PFDavg for the SIS sensors and logic solver is PFDavg = 0.1

Method

- Draw a fault tree for option 1 starting with top event = 'hazardous event'
- Split the causes of failure into two branches:
    - V1 and V2 both fail on demand from process
    - V1 fails on demand from trip demand caused by failure of V2
- Develop both branches downwards to show the fault logic
- Insert the faults/yr and PFD data on the fault tree and derive the trip failure rate for the top event.
- Revise the diagram and figures to show option 2 and option 3.
- Compare the results and comment on the value of installing extra valves.

Information only
Equations used for PFDs: (not required for the exercise but can checked anyway!)
- For single valve: **PFDavg = λdd .Ti/2**

- For V1 +V2 in 1oo2 operation : $\mathbf{PFDavg} = \{1-\beta\}\{(\lambda dd)^2 . (Ti)^2\}/3 + \beta.\lambda dd .Ti/2$

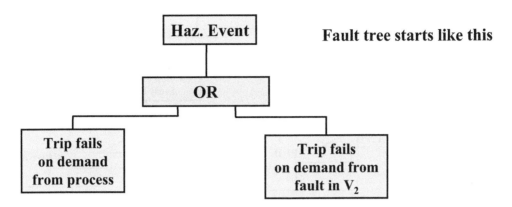

Fault tree starts like this

Answers to Exercise 6

- See fault tree diagram below for option 1:
  Hazardous event rate for option 1, V1 with V2 shared is 0.03/yr
- See fault tree diagram for option 2:
  Hazardous event rate for option 2, V2 only is 0.24/yr
- See fault tree diagram starting with option 3:
  Hazardous event rate for option 3, V1 as trip valve, is 0.072/yr

Conclusion:

- Option 2 is unacceptable. The link between valve failure causing a demand and the failure of the valve to protect the plant leads to a very poor risk reduction factor of 1.2/0.26 = 5 which is below SIL 1.
- Option 1 gives the best result. RRF = 1.2 /0.03 = 40 or SIL 1.
- Option 3 still meets SIL 1 but with a much lower RRF = 1.2/0.072 = 16.6.
- Both option 1 and 3 provide the essential element of independence but it is probably worth the additional cost of the solenoid trip device for V2 to have the better RRF of option 1.

Commentary: The fault tree diagrams show how we can break out the causes of a failure into separate functions and apply simple analysis to each part. The models shown here are useful for dealing with typical conceptual design issues.

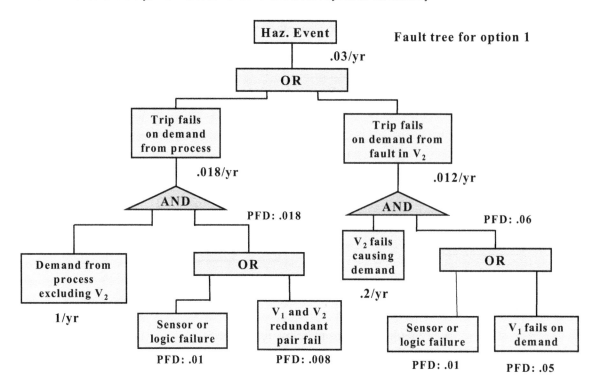

Fault tree for option 1

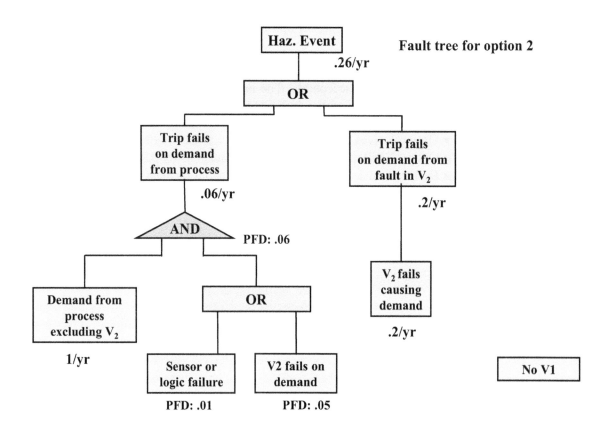

Fault tree for option 2

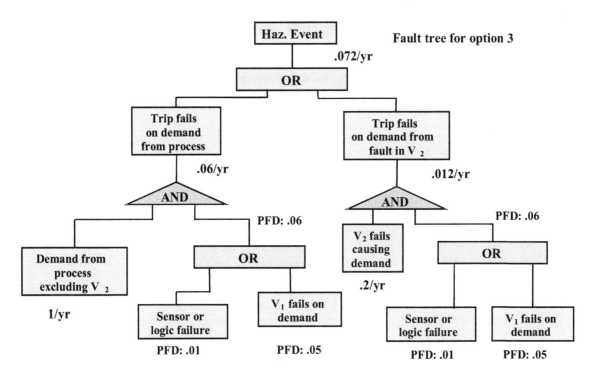

Fault tree for option 3

Exercise 7 – Exercise in conceptual design and in defining functional requirements

This is an exercise in evaluating the safety integrity of a proposed design for a trip system output stage where options have to be considered for a control valve to be used as a trip valve or as a redundant back up to a trip valve.

Subject	Defining the safety function
Objective	To experience practical methods for defining the safety function
Relevance	Chaper 4. Safety Requirements Spec. Chapter 7 Instruments Design
Task detail	1) Provide a conceptual design for a safety trip system that will protect the plant against defined risks during normal operation and during start up. 2) Define the functional requirements in a logic diagram suitable for documenting into a safety requirements specification.

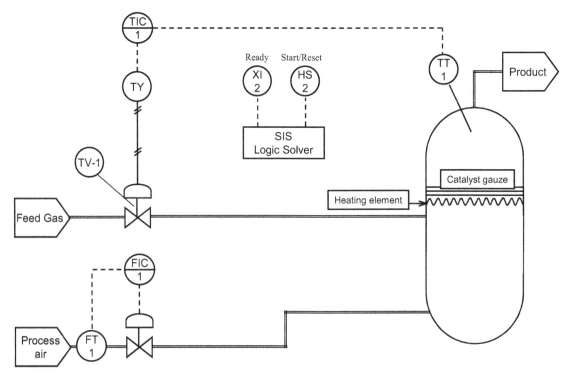

Practical 8 Fig 1 : Basic control scheme

Starting information

Refer to Figure 1; in this gas conversion process a feed gas mixture is passed over a hot catalyst and is converted to a stable product. An air/gas mixture is achieved by feeding air into the converter chamber under flow control. The reaction is exothermic and the catalyst becomes hotter as fuel flow is increased. If the gas temperature leaving the catalyst is less than 650° C there is a risk that an explosive mixture will be formed in the downstream plant. If the same temperature exceeds 750° C the catalyst will melt and the explosive mixture risk will again be present.

Temperature control of the process is achieved by a feedback loop to the flow control valve TV-1 from a temperature sensor TT-1 at the exit of the converter. At start up a minimum level of air flow is set up by the operator and electrical heating is applied to the catalyst gauze. Once operating temperature is achieved the feed gas can be introduced gradually. At first the catalyst temperature will fall slightly and then it will rise as the conversion process becomes established. About 1 minute after start up the temperature should stabilize and the loop TIC-1 can be put into Auto. Throughput is then increased by raising the air flow with the fuel feed following by temperature control.

Requirements for a safety instrumented system as defined from the hazard study are:

- Prevent opening of feed gas valve until minimum air flow has been present for more than 10 minutes and exit temperature is above 650° C and below 750° C.
- Provide an indicator (XI-2) that the trip system and the plant conditions are available for start up and provide a trip reset/start pushbutton (HS-2) for the operator to initiate the start up.
- During the first 60 seconds after start up the low temperature condition will be bypassed to allow a short temperature dip to occur.
- After 60 seconds both the lower and upper temperature limits will apply and any temperature excursion beyond these limits will cause the feed gas supply to be shut off and remain shut off until a new start up is required.
- The safe state of the process is when the feed gas valve is closed and air feed has been flowing above minimum for more than 10 minutes.
- Risk reduction requirements call for a safety integrity level of 1 (SIL 1).

The practical problem here is to sketch in a suitable outline of the design good enough to verify that it is feasible and workable without going into too much detail. Design iterations can then be made quickly if it is not right the first time.

Once the proposed design is in place it will be necessary to create a functional logic diagram (FLD) that will exactly describe the functional requirements. It is not always easy to get this right the first time around but once it is correct it provides a secure method of recording the design intent into a safety requirements specification. Often this FLD is then easy to translate into the configuration FLD for the final product. This assists in the estimating of the application software task.

Method:
- Complete the conceptual design by marking in on the diagram suitable sensors and actuators for the SIS with dashed lines into the logic solver. (Assume a PES based logic solver.)
- Check that these devices provide all the data required by the logic solver to perform the safety function. Take into account any common cause failure possibilities define the fail-to-safety modes of each device.
- Start a functional logic diagram showing all inputs on the left and all outputs on the right.

- Draw in the logic of the functions using logic block elements limited to AND, OR, NOT and TIMER. The basic ANSI symbols are shown below for guidance, alternative symbols can be used if they are defined.
- Verify the result against the narrative description given above.

Information only:

Any logical symbol set may be used for defining functional logic. However the convention used must be strictly applied to be reliable. The following symbols are based on ANSI Y32 14-1973 'Graphic Symbols for Logic Diagrams'. These can be used for the FLD required in this practical.

ANSI symbols for functional logic elements.

1) AND Symbol   Truth table

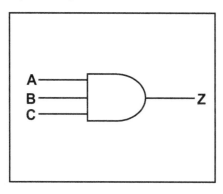

	Inputs		Outputs
A	B	C	Z
0	0	0	0
0	0	1	0
0	1	0	0
0	1	1	0
1	0	0	0
1	0	1	0
1	1	0	0
1	1	1	1

2) OR Symbol   Truth table

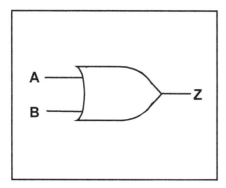

	Inputs	Outputs
A	B	Z
0	0	0
0	1	1
1	0	1
1	1	1

3) NOT Symbol   Truth table

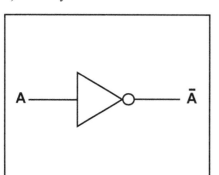

Input	Output
A	$\bar{A}$
0	1
1	0

4) Timer symbol; Delay initiation   Action

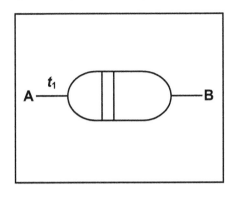

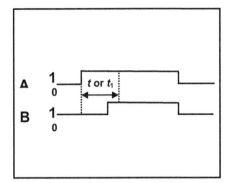

5) Timer symbol; 'Delay Termination'   Action

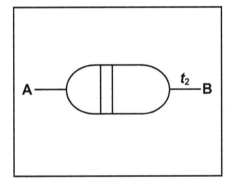

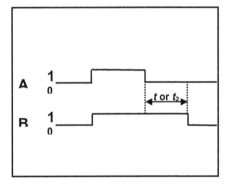

Answers to Exercise 7 – Conceptual design and in defining functional requirements

The first part of the practical requires that the SIS arrangement be drawn in on the diagram. The P&ID will then show the sensors and the actuator for the SIS functions.

The SIS will require independent sensors for temperature in the catalyst and for proving that air flow has been present for 10 minutes before the start up can be allowed. These are shown below as TT-2 and FT-2.

The actuator required to trip out the fuel feed should be fully independent of the control valve TV-1 since there is a high risk of common cause failure if TV-1 is used to shut off the fuel flow. The actuator is shown as TV-2 with a solenoid valve used to vent the air supply from the valve cylinder.

Fail-safe modes of the devices are shown by the arrow symbols of the actuators. For the sensors the design should specify upscale burnout modes for both temperature sensors since failures upscale would force the control system and the SIS into a safe mode for the process.

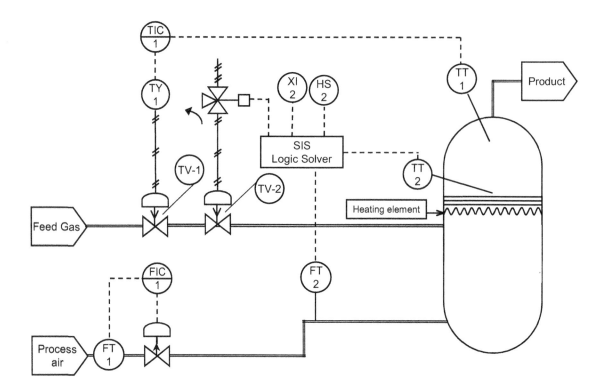

The functional logic diagram for the SIS must be arranged to show the complete operational cycle. The diagram shown below uses AND, OR and NOT function blocks with two timers to define the logical requirements.

The top section defines the temperature limiting conditions. The section below creates permissive conditions for the start up based on air flow greater than minimum for more than 10 minutes. This is combined with the temperature limits to operate the ready lamp. Finally the operator can press the start button and a 1-minute bypass condition on low temperature limit created. After this period the catalyst must be in the safe range of temperature at all times to allow the operation to continue. Any excursion outside of the limits will cause a trip, which will stay tripped until the 'ready' conditions are created again.

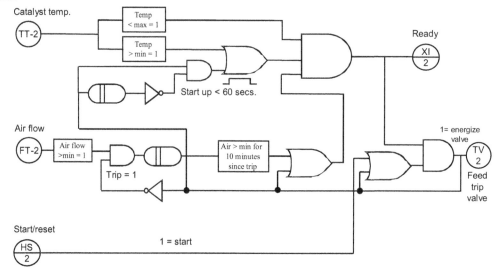

Appendix B

Glossary

Availability:

The probability that an item of equipment or a control system will perform its intended task. It is often expressed as a percentage of the time per year of use.

BPCS: basic process control system

Generic term used to describe any control system equipment provided for the normal operation of a plant or machine. A BPCS may or may not include safety functions.

CASS:

Conformity Assessment of Safety related Systems. Refers to the developing methods for assessment of project execution, equipment design as well as Functional Safety Management Capabilities. In the UK, accredited certification bodies will be available at a future date to offer CASS assessment services to industry.

Cause and effect diagram:

A matrix drawing showing the functional process safety interlocks between inputs and outputs of a safety system (see also 'FLD').

Common-cause failure:

Failure as a result of one or more events, originating from the same external or internal conditions, causing coincident failures of two or more separate channels in a multiple channel system (see also 'systematic failures').

Coverage factor: See Diagnostic coverage.

Covert failure:

A non-revealed defect in a system that is not detected by the incorporated test.

De-energized safe condition:

In this context: the electrical or pneumatic valves, which can shutdown the guarded process, are energized during the normal (safe) process situation. If an unsafe condition arises, the (spring-loaded) valve will close, because the energy is cut off.

Diagnostic coverage:

The efficacy of the self-diagnostics of a SIS, which makes it possible that a system successfully detects a specific type of component or software fault. IEC 61508 defines diagnostic coverage as 'fractional decrease in the probability of dangerous hardware failures resulting from the operation of the automatic diagnostic tests'.

Diagnostic coverage factor (also known as C-factor):

The C-factor comprises the percentage of failures in modules, software, external wiring, internal wiring, cables, interconnections and other functions that are detected by the built-in test functions, or by a suitable test program. It can be expressed in a probability or in a factor that is always smaller than 1 (e.g. $C = 0.95$) or as a percentage (e.g. 95%).

DCS:

Distributed (or Digital) control system. A process control system based on computer intelligence and using a data-highway to distribute the different functions to specialized controllers.

Dynamic logic circuit:

In this context: the valid logic-state can only exist and perform logic control, if the circuit is activated continuously, using alternating logic signals.

Emergency shutdown:

Commonly used terminology to refer to the safeguarding systems intended to shutdown a plant in case of a process parameter limit-excess. See also SRS and SIS.

EMI:

Electrical-magnetic interference.

EMC:

Electrical-magnetic compatibility.

E/E/PES (Electrical/electronic/programmable electronic system):

System for control, protection or monitoring based on one or more electrical/electronic programmable electronic (E/E/PE) devices, including all elements of the system such as power supplies, sensors and other input devices, data highways and other communication paths, and actuators and other output devices.

EUC: Equipment under control

Equipment, machinery, apparatus or plant used for manufacturing, process, transportation, medical or other activities.

EUC control system:

System which responds to input signals from the process and/or from an operator and generates output signals causing the EUC to operate in the desired manner.

NOTE – The EUC control system includes input devices and final elements. See also BPCS.

Fail-safe:

A control system response that, after one or multiple failures, lapses into a predictable safe condition.

Failure modes:

In safety-instrumented systems, 4 types of failure mode are recognized:

Detected safe failure, (revealed fault).

Undetected safe failure (an unrevealed fault that leads to a dangerous state).

Detected dangerous failure (a fault that is potentially dangerous but is detected by the system diagnostics, see revealed fault).

Undetected dangerous failure (a fault that prevents the system from providing its safety function and remains hidden within the system permanently or until found by periodic functional testing.

Fault:

IEC definition: *'abnormal condition that may cause a reduction in, or loss of, the capability of a functional unit to perform a required function'*.

Fault tolerance:

IEC definition: *'ability of a functional unit to perform a required function in presence of faults or errors'*.

FLD:

Functional logic diagram. A graphical representation of the system functions, showing the logic-gates and timers as well as the logic signal interconnections.

FMEA:

Failure mode and effect analysis. See IEC 61508 part 7 for description.

HIPPS:

High integrity pressure protection system. Also called 'over pressure protection system'.

HMI:

Human to machine interface or 'operator interface', usually a computer screen to present the actual process and system status.

IEC:

International Electrotechnical Commission. Based in Geneva. Develops a vast range of internationally supported standards. See website list.

Inherently fail-safe:

A particular designed dynamic logic principle that achieves the fail-safe property, from the principle itself and not from additional components or test circuits.

ISA: Instrument Society of America

Based in Research Triangle Park, North Carolina. Develops standards, technical reports and training material for the complete range of instrumentation with strong emphasis on process industries. See website list.

Logic solver:

E/E/PES components or subsystems that execute the application logic. Electronic and programmable electronics include input/output modules.

MTBF:

Mean time between failures. This term is normally applied to serviceable equipment, typically instrument sensors, valves or PLCs. Hence normally used in SIS reliability calculations.

MTTF:

Mean time to fail. This term is normally applied to disposable single life components such as relays or resistors which are replaced when they fail. Numerically the same as MTBF when calculating reliability of an SIS.

MTTR:

Mean time to repair. The mean time between the occurrence of a failure and the return to normal failure-free operation after a corrective action. This time also includes the time required for failure detection, failure search and re-starting the system.

Nuisance failure: See 'spurious trip'.

Overt faults:

Faults that are classified as announced, detected, revealed, etc. Opposite of 'covert fault'.

PFD:

Probability of failure on demand (PFD): The probability of a system failing to respond to a demand for action arising from a potentially hazardous condition. This parameter degrades (increases) during the mission time or test interval time. Therefore the average figure, PFDavg, is used in calculating the reliability of a safety system over a given mission time. PFD equals 1 minus safety availability.

PLC:

Programmable logic controller.

Proof test:

A 100% functional system test. In practice, this is only possible when the SIS is disconnected from the process. Hence on-line proof testing may leave a small fraction of the SIS untested. Also termed 'trip testing'.

Redundancy (identical and diverse):

Identical redundancy involves the use of elements identical in design, construction and in function with the objective to make the system more robust for self-revealing failures. 'Diverse redundancy' uses non-identical elements and provides a greater degree of protection against the potential for common cause faults. It can apply to hardware as well as to software.

Reliability:

The probability that no functional failure has occurred in a system during a given period of time.

Reliability block diagram:

The reliability block diagram can be thought of as a flow diagram from the input of the system to the output of the system. Each element of the system is a block in the reliability block diagram and, the blocks are placed in relation to the SIS architecture to indicate that a path from the input to the output is broken if one (or more) of the elements fail.

Revealed failure:

A failure in a system that results in a safe failure state of the system or is detected by the system's self-diagnostics. Also known as a safe detected failure.

Safety availability:

Probability that an SIS is able to perform its designated safety service when the process is operating. The average probability of failure on demand (PFDavg) is the preferred term. (PFD equals 1 minus safety availability.)

Safety instrumented systems (SIS):

System composed of sensors, logic solvers, and final control elements for the purpose of taking the process to a safe state when predetermined conditions are violated. Other terms commonly used include emergency shutdown system (ESD, ESS), safety shutdown system (SSD), and safety interlock system.

Safety life cycle:

Necessary activities involved in the implementation of safety related systems, occurring during a period of time that starts at the concept phase of a project and finishes when all of the E/E/PE safety related systems, other technology safety related systems and external risk reduction facilities are no longer available for use.

SCADA:

Supervisory control and data acquisition. This term is most commonly applied to PC based equipment interfaced to plant via PLCs or input–output devices.

SER:

Sequence of events recorder, based on real-time state changes of events in the system.

SIL:

Safety integrity level defining a PFD by order of magnitude, which is related to the risk (PFD) involved in various types of processes. In practice the SIL range is from 1 to 4 for most industrial processes.

Solid-state logic:

A term used to describe circuits whose functionality depends upon the interconnection of electronic components as semiconductors, resistors, capacitors, magnetic cores, etc and which do not depend on programmable electronics.

Spurious trip:

A plant trip arising out of an overt or detected equipment failure in the SIS or an erroneous assessment of the situation (e.g. error in the logic functions). A shutdown is initiated, though no real impairment of safety exists. Also referred to as a 'false trip' or a 'nuisance failure'. Spurious trips can contribute to the hazard rate of the plant through the disturbances so caused.

Systematic failures:

Failures occurring in identical parts of a (redundant) system due to similar circumstances. History shows that also errors in specification, engineering, software and environmental factors, such as electrical interference or maintenance errors must be considered. Such faults can only be eliminated by a modification of the design or of the manufacturing process, operational procedures, documentation or other relevant factors.

TMR: Triple modular redundancy

An architecture for SIS logic solvers to achieve fault-tolerance by a 2 out of 3 voting configuration using identical redundant modules.

Trip:

A shutdown of the process or machinery by a safety system. Normally a trip implies that the equipment cannot start operating again until there is a manually initiated restart procedure.

TÜV: Technische Üeberwachungs Verein

A testing laboratory in Germany that certifies safety of equipment in terms of compliance with international standards or German national standards.

Unrevealed failure:

A failure that impairs the system safety, but remains undetected (see also under 'covert failure'). It is related to the risk (PFD) involved in various types of processes. These types of failures can accumulate in a safety system, causing a degradation of the safety performance (SIL), as a function of time.

Index

Printed and bound by CPI Group (UK) Ltd, Croydon, CR0 4YY

03/10/2024

01040331-0018